ENERGY POLICIES, POLITICS AND PRICES

AMERICA'S PIPELINES

SAFETY AND SECURITY ISSUES

ENERGY POLICIES, POLITICS AND PRICES

Additional books in this series can be found on Nova's website under the Series tab.

Additional E-books in this series can be found on Nova's website under the E-book tab.

ENERGY POLICIES, POLITICS AND PRICES

AMERICA'S PIPELINES

SAFETY AND SECURITY ISSUES

JARON WIACEK
EDITOR

nova publishers
New York

For permission to use material from this book please contact us:
Telephone 631-231-7269; Fax 631-231-8175
Web Site: http://www.novapublishers.com

Library of Congress Cataloging-in-Publication Data

ISBN: 978-1-62417-994-5

Published by Nova Science Publishers, Inc. † New York

CONTENTS

PREFACE

Nearly half a million miles of high-volume pipeline transport natural gas, oil, and other hazardous liquids across the United States. These transmission pipelines are integral to U.S. energy supply and have vital links to other critical infrastructure, such as power plants, airports, and military bases. While an efficient and fundamentally safe means of transport, many pipelines carry volatile, flammable, or toxic materials with the potential to cause public injury and environmental damage. The nation's pipeline networks are also widespread, running alternately through remote and densely populated regions, some above ground, some below; consequently, these systems are vulnerable to accidents and terrorist attack. This book provides an overview of the federal pipeline safety program and the federal role in pipeline security, with a focus on the nation's overall strategy to protect transportation infrastructure.

Chapter 1 – Nearly half a million miles of pipeline transporting natural gas, oil, and other hazardous liquids crisscross the United States. While an efficient and fundamentally safe means of transport, many pipelines carry materials with the potential to cause public injury and environmental damage. The nation’s pipeline networks are also widespread and vulnerable to accidents and terrorist attack. Recent pipeline accidents in Marshall, MI, San Bruno, CA, Allentown, PA, and Laurel, MT, have heightened congressional concern about pipeline risks and drawn criticism from the National Transportation Safety Board. Both government and industry have taken numerous steps to improve pipeline safety and security over the last 10 years. Nonetheless, while many stakeholders agree that federal pipeline safety programs have been on the right track, the spate of recent pipeline incidents suggest there continues to be significant room for improvement. Likewise, the threat of terrorist attack remains a concern.

The federal pipeline safety program is authorized through the fiscal year ending September 30, 2015, under the Pipeline Safety, Regulatory Certainty, and Job Creation Act of 2011 (P.L. 112-90) which was signed by President Obama on January 3, 2012. The act contains a broad range of provisions addressing pipeline safety and security. Among the most significant are provisions that could increase the number of federal pipeline safety inspectors, require automatic shutoff valves for transmission pipelines, mandate verification of maximum allowable operating pressure for gas transmission pipelines, increase civil penalties for pipeline safety violations, and mandate reviews of diluted bitumen pipeline regulation. The Transportation Security Administration Authorization Act of 2011 (H.R. 3011) would mandate a study regarding the relative roles and responsibilities of the Department of Homeland Security and the Department of Transportation with respect to pipeline security.

As it oversees the federal pipeline safety program and the federal role in pipeline security, Congress may wish to assess how the various elements of U.S. pipeline safety and security fit together in the nation's overall strategy to protect transportation infrastructure. Pipeline safety and security necessarily involve many groups: federal agencies, oil and gas pipeline associations, large and small pipeline operators, and local communities. Reviewing how these groups work together to achieve common goals could be an oversight challenge for Congress.

Chapter 2 – The vast U.S. network of natural gas and hazardous liquid pipelines is integral to U.S. energy supply and has vital links to other critical infrastructure. While an efficient and fundamentally safe means of transport, this network is vulnerable to cyber attacks. In particular, cyber infiltration of supervisory control and data acquisition (SCADA) systems could allow successful "hackers" to disrupt pipeline service and cause spills, explosions, or fires—all from remote locations. In March 2012, the Department of Homeland Security (DHS) reported ongoing cyber intrusions among U.S. natural gas pipeline operators. These intrusions have heightened congressional concern about cybersecurity in the U.S. pipelines sector.

The Transportation Security Administration (TSA) is authorized by federal statute to promulgate pipeline physical security and cybersecurity regulations, if necessary, but the agency has not issued such regulations. TSA officials assert that security regulations could be counterproductive because they could establish a general standard below the level of security already in place for many pipelines. An April 2011 White House proposal and the Cybersecurity Act of 2012 (S. 2105) both would mandate cybersecurity regulations for privately owned critical infrastructures sectors like pipelines. A revised version of S. 2105, S. 3414, would permit the issuance of regulations but would focus on voluntary cybersecurity measures.

Chapter 3 – Pipelines are a relatively safe mode of transportation for hazardous liquid and natural gas and are regulated by the Department of Transportation's (DOT) Pipeline and Hazardous Materials Safety Administration (PHMSA) and state entities. Included in the nation's pipeline network are an estimated 200,000 or more miles of onshore "gathering" pipelines, which transport products to processing facilities and larger pipelines. (See figure.) Many of these pipelines have not been subject to federal regulation based on their generally rural location and low operating pressures. While incidents involving gathering pipelines regulated by PHMSA have resulted in millions of dollars in property damage in recent years, comparable statistics for federally unregulated gathering pipelines are unknown. This report identifies (1) the safety risks that exist, if any, with onshore hazardous liquid and natural gas gathering pipelines that are not currently under PHMSA regulation and (2) the practices states use to help ensure the safety of these pipelines. GAO surveyed state pipeline safety agencies in all 50 states and the District of Columbia; interviewed officials at PHMSA, state pipeline safety agencies, pipeline companies, and industry associations; and analyzed data and regulations.

Chapter 4 – Applying PHMSA's new distribution integrity management requirements to low-stress transmission pipelines would result in less prescriptive safety requirements for these pipelines. Overall, requirements for distribution pipelines are less prescriptive than requirements for transmission pipelines in part because the former operate at lower pressure and pose lower risks in general than the latter. For example, the integrity management regulations for transmission pipelines allow three types of in-depth physical inspections.4 In contrast, distribution pipeline operators can customize their integrity management programs to the complexity of their systems, including using a broader range of methods for physical inspection. While PHMSA officials stated that "less prescriptive" does not necessarily mean less safe, they also stated that distribution integrity management requirements for distribution pipelines can be more difficult to enforce than integrity management requirements for transmission pipelines. Currently, PHMSA can grant special permits to modify requirements for individual pipelines, if merited,5 but applying the new distribution integrity management requirements to low-stress transmission pipelines would affect all such pipelines.

In: America's Pipelines
Editor: Jaron Wiacek
ISBN: 978-1-62417-994-5

Chapter 1

KEEPING AMERICA'S PIPELINES SAFE AND SECURE: KEY ISSUES FOR CONGRESS*

Paul W. Parfomak

SUMMARY

Nearly half a million miles of pipeline transporting natural gas, oil, and other hazardous liquids crisscross the United States. While an efficient and fundamentally safe means of transport, many pipelines carry materials with the potential to cause public injury and environmental damage. The nation's pipeline networks are also widespread and vulnerable to accidents and terrorist attack. Recent pipeline accidents in Marshall, MI, San Bruno, CA, Allentown, PA, and Laurel, MT, have heightened congressional concern about pipeline risks and drawn criticism from the National Transportation Safety Board. Both government and industry have taken numerous steps to improve pipeline safety and security over the last 10 years. Nonetheless, while many stakeholders agree that federal pipeline safety programs have been on the right track, the spate of recent pipeline incidents suggest there continues to be significant room for improvement. Likewise, the threat of terrorist attack remains a concern.

The federal pipeline safety program is authorized through the fiscal year ending September 30, 2015, under the Pipeline Safety, Regulatory

* This is an edited, reformatted and augmented version of a Congressional Research Service publication, CRS Report for Congress R41536, prepared for Members and Committees of Congress, from www.crs.gov, dated March 13, 2012.

Certainty, and Job Creation Act of 2011 (P.L. 112-90) which was signed by President Obama on January 3, 2012. The act contains a broad range of provisions addressing pipeline safety and security. Among the most significant are provisions that could increase the number of federal pipeline safety inspectors, require automatic shutoff valves for transmission pipelines, mandate verification of maximum allowable operating pressure for gas transmission pipelines, increase civil penalties for pipeline safety violations, and mandate reviews of diluted bitumen pipeline regulation. The Transportation Security Administration Authorization Act of 2011 (H.R. 3011) would mandate a study regarding the relative roles and responsibilities of the Department of Homeland Security and the Department of Transportation with respect to pipeline security.

As it oversees the federal pipeline safety program and the federal role in pipeline security, Congress may wish to assess how the various elements of U.S. pipeline safety and security fit together in the nation's overall strategy to protect transportation infrastructure. Pipeline safety and security necessarily involve many groups: federal agencies, oil and gas pipeline associations, large and small pipeline operators, and local communities. Reviewing how these groups work together to achieve common goals could be an oversight challenge for Congress.

INTRODUCTION

Nearly half a million miles of high-volume pipeline transport natural gas, oil, and other hazardous liquids across the United States.[1] These transmission pipelines are integral to U.S. energy supply and have vital links to other critical infrastructure, such as power plants, airports, and military bases. While an efficient and fundamentally safe means of transport, many pipelines carry volatile, flammable, or toxic materials with the potential to cause public injury and environmental damage. The nation's pipeline networks are also widespread, running alternately through remote and densely populated regions, some above ground, some below; consequently, these systems are vulnerable to accidents and terrorist attack. Recent pipeline accidents in Marshall, MI, San Bruno, CA, Allentown, PA, and Laurel, MT, have demonstrated this vulnerability and have heightened congressional concern about pipeline risks.

The federal program for pipeline safety resides primarily within the Department of Transportation (DOT), although its inspection and enforcement activities rely heavily upon partnerships with state pipeline safety agencies. The federal pipeline security program began with the DOT as well, immediately after the terror attacks of September 11, 2001, but pipeline security authority was subsequently transferred to the Department of Homeland Security (DHS) when the latter department was created. The DOT and DHS have distinct missions, but they cooperate to protect the nation's pipelines. The Federal Energy Regulatory Commission is not operationally involved in pipeline safety or security, but it can examine safety issues under its siting authority for interstate natural gas pipelines, and can allow pipeline companies under its rate jurisdiction to recover pipeline security costs. Collectively, these agencies administer a comprehensive and complex set of regulatory authorities which has been changing significantly over the last decade and continues to do so.

The federal pipeline safety program is authorized through the fiscal year ending September 30, 2015, under the Pipeline Safety, Regulatory Certainty, and Job Creation Act of 2011 (P.L. 112-90) which was signed by President Obama on January 3, 2012. This report reviews the history of federal programs for pipeline safety and security, key policy issues, and recent developments leading up to P.L. 112-90. Although the act contains over 30 substantive sections, this report focuses only on a subset of provisions of key interest in recent congressional debate.

SAFETY AND SECURITY IN THE PIPELINE INDUSTRY

Of the nation's approximately half million miles of transmission pipeline, roughly 170,000 miles carry hazardous liquids—over 75% of the nation's crude oil and around 60% of its refined petroleum products, along with other products.[2] Within this network, there are nearly 200 *inter*state crude oil and liquid fuel pipelines, which account for roughly 80% of total pipeline mileage and transported volume.[3]

The U.S. natural gas pipeline network consists of around 217,000 miles of *inter*state transmission, and 89,000 miles of *intra*state transmission.[4] It also contains some 20,000 miles of field and gathering pipeline, which connect gas extraction wells to processing facilities.[5] Around 120 systems make up the *inter*state gas transmission network; another 90 or so systems operate strictly within individual states.[6] These *inter*state and *intra*state gas transmission pipelines feed around 1.2 million miles of regional pipelines in some 1,400 local distribution networks.[7] Natural gas pipelines also connect to 113 liquefied natural gas (LNG) storage sites, which augment pipeline gas supplies during peak demand periods.[8]

Pipeline Safety Record

Taken as a whole, releases from pipelines cause few annual fatalities compared to other product transportation modes. According to the DOT, hazardous liquid pipelines reported an average of 1.8 deaths per year from 2006 through 2010. During the same period, natural gas transmission and distribution pipelines reported an average of 3.0 and 9.8 deaths per year, respectively.[9] Accidental pipeline releases result from a variety of causes, including third-party excavation, corrosion, mechanical failure, control system failure, and operator error. Natural forces, such as floods and earthquakes, can also damage pipelines. There were 120 hazardous liquid pipeline accidents, 81 natural gas transmission (including gathering) pipeline accidents, and 54 natural gas distribution accidents in 2010.[10]

Although pipeline releases have caused relatively few fatalities in absolute numbers, a single pipeline accident can be catastrophic in terms of deaths and environmental damage. Notable pipeline accidents in recent years include:

- *1999*—A gasoline pipeline explosion in Bellingham, WA, killed three people and caused $45 million in damage to a city water plant and other property.
- *2000*—A natural gas pipeline explosion near Carlsbad, NM, killed 12 campers.
- *2006*—Corroded pipelines on the North Slope of Alaska leaked over 200,000 gallons of crude oil in an environmentally sensitive area and temporarily shut down Prudhoe Bay oil production.
- *2007*—An accidental release from a propane pipeline and subsequent fire near Carmichael, MS, killed two people, injured several others, destroyed four homes, and burned over 70 acres of grassland and woodland.
- *2010*—A pipeline spill in Marshall, MI, released 819,000 gallons of crude oil into a tributary of the Kalamazoo River.
- *2010*—A natural gas pipeline explosion in San Bruno, CA, killed 8 people, injured 60 others, and destroyed 37 homes.
- *2011*—A natural gas pipeline explosion in Allentown, PA, killed 5 people, damaged 50 buildings, and caused 500 people to be evacuated.
- *2011*—A pipeline spill near Laurel, MT, released an estimated 42,000 gallons of crude oil into the Yellowstone River.

Such accidents have generated persistent scrutiny of pipeline regulation and have increased state and community activity related to pipeline safety.

Pipeline Security Risks

In addition to their vulnerability to accidents, pipelines may also be intentionally damaged by vandals and terrorists. Some pipelines may also be vulnerable to "cyber-attacks" on computer control systems or attacks on electricity grids and telecommunications networks.[11] Oil and gas pipelines, globally, have been a favored target of terrorists, militant groups, and organized crime. In Colombia, for example, rebels have bombed the Caño Limón oil pipeline and other pipelines over 950 times since 1993.[12] In 1996, London police foiled a plot by the Irish Republican Army to bomb gas pipelines and other utilities across the city.[13] Militants in Nigeria have repeatedly attacked pipelines and related facilities, including the simultaneous bombing of three oil pipelines in May 2007.[14] A Mexican rebel group similarly detonated bombs along Mexican oil and natural gas pipelines in July and September 2007.[15] In June 2007, the U.S. Department of Justice arrested members of a terrorist group planning to attack jet fuel pipelines and storage tanks at the John F. Kennedy (JFK) International Airport in New York.[16] Natural gas pipelines in British Columbia, Canada, were bombed six times between October 2008 and July 2009 by unknown perpetrators.[17]

In 2009, the *Washington Post* reported that over $1 billion of crude oil had been stolen directly from Mexican pipelines by organized criminals and drug cartels.[18]

Since September 11, 2001, federal warnings about Al Qaeda have mentioned pipelines specifically as potential terror targets in the United States.[19] One U.S. pipeline of particular concern, and with a history of terrorist and vandal activity, is the Trans Alaska Pipeline System (TAPS), which transports crude oil from Alaska's North Slope oil fields to the marine terminal in Valdez. TAPS runs some 800 miles and delivers nearly 17% of U.S. domestic oil production.[20] In 1999, Vancouver police arrested a man planning to blow up TAPS for personal profit in oil futures.[21] In 2001, a vandal's attack on TAPS with a high-powered rifle forced a two-day shutdown and caused extensive economic and ecological damage.[22] In January 2006, federal authorities acknowledged the discovery of a detailed posting on a website purportedly linked to Al Qaeda that reportedly encouraged attacks on U.S. pipelines, especially TAPS, using weapons or hidden explosives.[23] In

November 2007 a U.S. citizen was convicted of trying to conspire with Al Qaeda to attack TAPS and a major natural gas pipeline in the eastern United States.[24]

Notwithstanding the incidents cited above, to date, there have been no known Al Qaeda attacks on TAPS or other U.S. pipelines. The most recent U.S. federal threat assessment concludes "with high confidence that the terrorist threat to the U.S. pipeline industry is low ... [with] no specific or credible threat information indicating that violent transnational extremist groups or domestic extremists are actively plotting to conduct attacks on the U.S. pipeline industry."[25] Terrorist activities are in constant flux, however, and difficult to predict, so terrorist attacks remain a possibility in the future. Attacks by individuals not associated with organized groups also remain a concern, as demonstrated by the August 2011 arrest by federal agents of a U.S. citizen, acting alone, who confessed to planting an improvised explosive device (which failed to detonate) under a natural gas pipeline in Oklahoma.[26]

Pipelines and Hazardous Materials Safety Administration

The Natural Gas Pipeline Safety Act of 1968 (P.L. 90-481) and the Hazardous Liquid Pipeline Act of 1979 (P.L. 96-129) are two of the principal early acts establishing the federal role in pipeline safety. Under both statutes, the Transportation Secretary is given primary authority to regulate key aspects of interstate pipeline safety: design, construction, operation and maintenance, and spill response planning. Pipeline safety regulations are covered in Title 49 of the *Code of Federal Regulations*.[27] The DOT administers pipeline regulations through the Office of Pipeline Safety (OPS) within the Pipelines and Hazardous Materials Safety Administration (PHMSA). In FY2010 the OPS was funded for 206 full-time equivalent staff based in Washington, DC; Atlanta; Kansas City; Houston; and Denver.[28] This included funding for 135 inspectors, although the agency actually employed 110 inspectors at the end of FY2010.[29] In 2011, the agency was operating under a continuing resolution at the same funding levels. P.L. 112-90 contains provisions authorizing up to 145 inspection and enforcement personnel if necessary (§ 31(b)). In addition to its own staff, PHMSA's enabling legislation allows the agency to delegate authority to *intra*state pipeline safety offices, and allows state offices to act as "agents" administering *inter*state pipeline safety programs (excluding enforcement) for those sections of *inter*state pipelines within their boundaries.[30] Over 400 state pipeline safety inspectors are available in 2012.

PHMSA's pipeline safety program is funded primarily by user fees assessed on a per-mile basis on each regulated pipeline operator (49 U.S.C. §60107). P.L. 109-468 authorized annual pipeline safety program expenditures of $79.0 million in FY2007, $86.2 million in FY2008, $91.5 million in FY2009, and $96.5 million in FY2010. P.L. 112-90 authorizes expenditures of $109.3 million annually for each of FY2012 through FY2015, and $1.5 million annually through FY2015 for state pipeline damage prevention programs. The President's FY2012 budget request included $176.2 million, including funds for emergency preparedness grants.[31]

PHMSA uses a variety of strategies to promote compliance with its safety standards. The agency conducts programmatic inspections of management systems, procedures, and processes; conducts physical inspections of facilities and construction projects; investigates safety incidents; and maintains a dialogue with pipeline operators. The agency clarifies its regulatory expectations through published protocols and regulatory orders, guidance manuals, and public meetings. PHMSA relies upon a range of enforcement actions, including administrative actions such as corrective action orders (CAOs) and civil penalties, to ensure that operators correct safety violations and take measures to preclude future safety problems. In 2011, PHMSA initiated 237 enforcement actions against pipeline operators.[32] Civil penalties proposed by PHMSA for safety violations during this period totaled approximately $3.7 million.[33] PHMSA also conducts accident investigations and system-wide reviews focusing on high-risk operational or procedural problems and areas of the pipeline near sensitive environmental areas, high-density populations, or navigable waters.

Since 1997, PHMSA has increasingly required industry's implementation of "integrity management" programs on pipeline segments near "high consequence areas." Integrity management provides for continual evaluation of pipeline condition; assessment of risks to the pipeline; inspection or testing; data analysis; and follow-up repair, as well as preventive or mitigative actions. High consequence areas include population centers, commercially navigable waters, and environmentally sensitive areas, such as drinking water supplies or ecological reserves. The integrity management approach directs priority resources to locations of highest consequence rather than applying uniform treatment to the entire pipeline network. PHMSA made integrity management programs mandatory for most oil pipeline operators with 500 or more miles of regulated pipeline as of March 31, 2001 (49 C.F.R. §195).

DOT Pipeline Security Activities

Presidential Decision Directive 63 (PDD-63), issued during the Clinton Administration, assigned lead responsibility for pipeline security to the DOT.[34] These responsibilities fell to the OPS, at that time a part of the DOT's Research and Special Programs Administration (RSPA), since the agency was already addressing some elements of pipeline security in its role as safety regulator.[35] In 2002, the OPS conducted a vulnerability assessment to identify critical pipeline facilities and worked with industry groups and state pipeline safety organizations "to assess the industry's readiness to prepare for, withstand and respond to a terrorist attack."[36] Together with the Department of Energy and state pipeline agencies, the OPS promoted the development of consensus standards for security measures tiered to correspond with the five levels of threat warnings issued by the Office of Homeland Security.[37] The OPS also developed protocols for inspections of critical facilities to ensure that operators implemented appropriate security practices. To convey emergency information and warnings, the OPS established communication links to key staff at the most critical pipeline facilities throughout the country. The OPS also began identifying near-term technology to enhance deterrence, detection, response, and recovery, and began seeking to advance public and private sector planning for response and recovery.[38]

On September 5, 2002, the OPS circulated guidance developed in cooperation with the pipeline industry defining the agency's security program recommendations and implementation expectations. This guidance recommended that operators identify critical facilities, develop security plans consistent with prior trade association security guidance, implement these plans, and review them annually.[39] Although the guidance was voluntary, the OPS expected compliance and informed operators of its intent to begin reviewing security programs within 12 months, potentially as part of more comprehensive safety inspections.[40] Federal pipeline security authority was subsequently transferred outside of DOT, however, as discussed below, so the OPS did not follow through on a national program of pipeline security program reviews.

Pipeline Safety Improvement Act of 2002

On December 12, 2002, President Bush signed into law the Pipeline Safety Improvement Act of 2002 (P.L. 107-355). The act strengthened federal pipeline safety programs, state oversight of pipeline operators, and public education regarding pipeline safety.[41] Among other provisions, P.L. 107-355 required operators of regulated natural gas pipelines in high-consequence areas to conduct risk analysis and implement integrity management programs similar to those required for oil pipelines.[42] The act authorized the DOT to order safety actions for pipelines with potential safety problems and increased violation penalties. The act streamlined the permitting process for emergency pipeline restoration by establishing an interagency committee, including the DOT, the Environmental Protection Agency, the Bureau of Land Management, the Federal Energy Regulatory Commission, and other agencies, to ensure coordinated review and permitting of pipeline repairs. The act required DOT to study ways to limit pipeline safety risks from population encroachment and ways to preserve environmental resources in pipeline rights-of-way. P.L. 107-355 also included provisions for public education, grants for community pipeline safety studies, "whistle blower" and other employee protection, employee qualification programs, and mapping data submission.

Pipeline Inspection, Protection, Enforcement, and Safety Act of 2006

On December 29, 2006, President Bush signed into law the Pipeline Inspection, Protection, Enforcement and Safety Act of 2006 (PIPES Act, P.L. 109-468). The main provisions of the act address pipeline damage prevention, integrity management, corrosion control, and enforcement transparency. The PIPES act created a national focus on pipeline damage prevention through grants to states for improving damage prevention programs, establishing 811 as national "call before you dig" one-call telephone number, and giving PHMSA limited "backstop" authority to conduct civil enforcement against one-call violators in states that have failed to conduct such enforcement. The act mandated the promulgation by PHMSA of minimum standards for integrity management programs for natural gas distribution pipelines.[43] It also mandated a review of the adequacy of federal pipeline safety regulations related to internal corrosion control, and required PHMSA to increase the transparency of enforcement actions by issuing monthly summaries, including violation and penalty information, and a mechanism for pipeline operators to make response information available to the public.

Transportation Security Administration

In November 2001, President Bush signed the Aviation and Transportation Security Act (P.L. 107- 71) establishing the Transportation Security Administration (TSA) within the DOT. According to TSA, the act placed the DOT's pipeline security authority (under PDD-63) within TSA. The act specified for TSA a range of duties and powers related to general transportation security, such as intelligence management, threat assessment, mitigation, security measure oversight and enforcement, among others. On November 25, 2002, President Bush signed the Homeland Security Act of 2002 (P.L. 107-296) creating the Department of Homeland Security (DHS). Among other provisions, the act transferred to DHS the Transportation Security Administration from the DOT (§403). On December 17, 2003, President Bush issued Homeland Security Presidential Directive 7 (HSPD-7), clarifying executive agency responsibilities for identifying, prioritizing, and protecting critical infrastructure.[44] HSPD-7 maintains DHS as the lead agency for pipeline security (par. 15), and instructs the DOT to "collaborate in regulating the transportation of hazardous materials by all modes (including pipelines)" (par. 22h). The order requires that DHS and other federal agencies collaborate with "appropriate private sector entities" in sharing information and protecting critical infrastructure (par. 25). TSA joined both the Energy Government Coordinating Council and the Transportation Government Coordinating Council under provisions in HSPD-7. The missions of the councils are to work with their industry counterparts to coordinate critical infrastructure protection programs in the energy and transportation sectors, respectively, and to facilitate the sharing of security information.

HSPD-7 also required DHS to develop a national plan for critical infrastructure and key resources protection (par. 27), which the agency issued in 2006 as the *National Infrastructure Protection Plan* (NIPP). The NIPP, in turn, required each critical infrastructure sector to develop a Sector Specific Plan (SSP) that describes strategies to protect its critical infrastructure, outlines a coordinated approach to strengthen its security efforts, and determines appropriate funding for these activities. Executive Order 13416 further required the transportation sector SSP to prepare annexes for each mode of surface transportation.[45] In accordance with the above requirements the TSA issued its *Transportation Systems Sector Specific Plan* and *Pipeline Modal Annex* in 2007 with an update on 2010.

TSA Pipeline Security Activities

Pipeline security activities at TSA are led by the Pipeline Security Division (PSD) within the agency's Office of Security Policy and Industry Engagement (OSPIE).[46]

According to the agency's *Pipeline Modal Annex* (PMA), TSA has been engaged in a number of specific pipeline security initiatives as summarized in *Table 1.*

Table 1. Ongoing TSA Pipeline Security Initiatives

Initiative	Description	Participants
Pipeline System Relative Risk Tool	Statistical tool used for relative risk ranking and prioritizing CSR findings	TSA, Industry
Corporate Security Reviews (CSR)	On-site reviews of pipeline operator security	TSA, Industry
Critical Facility Inspections (CFI)	On-site inspection of critical facilities of the 100 most critical pipeline systems	TSA, Industry
Pipeline Security Guidance	Maintains and revises voluntary pipeline security guidelines and best practices with input from government and industry partners	TSA, Other agencies, Industry,
Security Incident and Recovery	Completed a security incident and recovery protocol plan mandated under P.L. 110-53	Industry
Toxic Inhalation Hazard Materials Program	Program to address potential risks from pipeline transportation of hazardous materials other than oil and natural gas	TSA, Other agencies, Industry
Pipeline Cross-Border Vulnerability Assessment	U.S. and Canadian security assessment and planning for critical cross-border pipeline	TSA, Canada
International Pipeline Security Forums	International forums for U.S. and Canadian governments and pipeline industry officials convened annually	TSA, Canada, Other agencies, Industry
Pipeline Security Exercises	Facilitation of pipeline security drills and exercises	TSA, Other Agencies,
	including those under the Intermodal Security Training Exercise Program (I-STEP)	Industry
Security Awareness Training Materials	Informational compact discs about pipeline security issues and improvised explosive devices	TSA
Stakeholder Conference Calls	Periodic information-sharing conference calls between key pipeline security stakeholders	TSA, Other agencies, Industry
Sector Coordinating Councils and Joint Sector Committee	Government partners coordinate interagency and cross- jurisdictional implementation of critical infrastructure security	TSA, Other agencies, Industry

Sources: Transportation Security Administration, Transportation Systems Sector-Specific Plan, 2010, pp. 326.

In addition to the activities in *Table 1*, TSA has also conducted regional supply studies for key natural gas markets, has conducted training on cyber security awareness, has participated in pipeline blast mitigation studies, and has joined in "G-8" multinational security assessment and planning.[47]

In 2003, TSA initiated its Corporate Security Review (CSR) program, wherein the agency visits the largest pipeline and natural gas distribution operators to review their security plans and inspect their facilities. During the reviews, TSA evaluates whether each company is following the intent of the OPS security guidance, and seeks to collect the list of assets each company had identified meeting the criteria established for critical facilities. In 2004, the DOT reported that the plans reviewed to date (approximately 25) had been "judged responsive to the OPS guidance."[48]

TSA has completed CSRs covering the largest 100 pipeline systems (84% of total U.S. energy pipeline throughput) and is in the process of conducting second CSRs of these systems.[49] According to TSA, CSR results indicate that the majority of U.S. pipeline systems "continue to do a good job in regards to pipeline security" although there are areas in which pipeline security can be improved.[50] Past CSR reviews have identified inadequacies in some company security programs such as not updating security plans, lack of management support, poor employee involvement, inadequate threat intelligence, and employee apathy or error.[51] In 2008, the TSA initiated its Critical Facility Inspection Program (CFI), under which the agency conducted in-depth inspections of all the critical facilities of the 125 largest pipeline systems in the United States. The agency estimated that these 125 pipeline systems collectively included approximately 600 distinct critical facilities.[52] TSA concluded the CFI program in May 2011, having completed a total of 347 inspections throughout the United States.[53]

In addition to the initiatives in *Table 1,* TSA has worked to establish qualifications for personnel applying for positions with unrestricted access to critical pipeline assets and has developed its own inventory of critical pipeline infrastructure.[54] The agency has also addressed legal issues regarding recovery from terrorist attacks, such as FBI control of crime scenes and eminent domain in pipeline restoration. In October 2005, TSA issued an overview of recommended security practices for pipeline operators "for informational purposes only ... not intended to replace security measures already implemented by individual companies."[55] The agency released revised pipeline security guidelines in 2010 and 2011. The guidelines include a section on cybersecurity developed with the assistance of the Applied Physics Laboratory of Johns Hopkins University as well as other government and industry stakeholders.[56]

The President's FY2012 budget request for DHS did not include a separate line item for TSA's pipeline security activities. The budget request did include a $134.7 million line item for "Surface Transportation Security," which encompasses security activities in non-aviation transportation modes, including pipelines.[57]

The PSD has traditionally received from the agency's general operational budget an allocation for routine operations such as regulation development, travel, and outreach. According to the PSD, the budget funds 13 full-time equivalent staff within the office.[58]

In 2007 the TSA Administrator testified before Congress that the agency intended to conduct a pipeline infrastructure study to identify the "highest risk" pipeline assets, building upon such a list developed through the CSR program. He also stated that the agency would use its ongoing security review process to determine the future implementation of baseline risk standards against which to set measurable pipeline risk reduction targets.[59]

Provisions in the Implementing Recommendations of the 9/11 Commission Act of 2007 (P.L. 110-53) require TSA, in consultation with PHMSA, to develop a plan for the federal government to provide increased security support to the "most critical" pipelines at high or severe security alert levels and when there is specific security threat information relating to such pipeline infrastructure (§1558(a)(1)).

The act also requires a recovery protocol plan in the event of an incident affecting the interstate and intrastate pipeline system (§1558(a)(2)). TSA published this plan in 2010.

Security Incident Investigations

In addition to the above pipeline security initiatives, the TSA Pipeline Security Division has performed a limited number of vulnerability assessments and has supported investigations for specific companies and assets where intelligence information has suggested potential terrorist activity.

The PSD, along with PHMSA, was involved in the investigation of an August 2006 security breach at an LNG peak-shaving plant in Lynn, MA.[60] Although not a terrorist incident, the security breach involved the penetration of intruders through several security barriers and alert systems, permitting them to access the main LNG storage tank at the facility.

The PSD also became aware of the JFK airport terrorist plot in its early stages and supported the Federal Bureau of Investigation's associated investigation.

The PSD engaged the private sector in helping to assess potential targets and determine potential consequences. The PSD worked with the pipeline company to keep it informed about the plot, discuss its security practices, and review its emergency response plans.[61]

GAO Study of TSA's Pipeline Security Activities

In December 2008, the Senate Committee on Commerce, Science, and Transportation requested a study by the Government Accountability Office (GAO) examining TSA's efforts to ensure pipeline security. GAO's report, released in August 2010, focused on TSA's use of risk assessment and risk information in securing pipelines, actions the agency has taken to improve pipeline security under guidance in the 9/11 Commission Act of 2007 (P.L. 110-53), and the agency's efforts to measure such security improvement efforts.[62] Among other findings, GAO concluded that, although TSA had begun to implement a risk management approach to prioritize its pipeline security efforts, work remained to ensure that the highest risk pipeline systems would get the necessary scrutiny. GAO also concluded that TSA was missing opportunities under its CSR and CFI programs to better ensure that pipeline operators understand how they can enhance the security of their pipeline systems. TSA could also make better use of CSR and CFI recommendations for analyzing pipeline vulnerabilities and was not following up on these recommendations. GAO found that linking TSA's pipeline security performance measures and milestones to the goals and objectives in its national security strategy for pipeline systems could aid in achieving results within specific time frames and could facilitate more effective oversight and accountability.[63] TSA concurred with all of GAO's recommendations for addressing the issues and has since been implementing them.[64]

Federal Energy Regulatory Commission

One area related to pipeline safety and security not under either PHMSA's or TSA's primary jurisdiction is the siting approval of new gas pipelines, which is the responsibility of the Federal Energy Regulatory Commission (FERC). Companies building interstate natural gas pipelines must first obtain from FERC certificates of public convenience and necessity. (FERC does not oversee oil pipeline construction.) FERC must also approve the abandonment of gas facility use and services. These approvals may include safety and security provisions with respect to pipeline routing, safety standards and other factors.[65] As a practical matter, however, FERC has traditionally left these considerations to the other agencies.[66]

On September 14, 2001, FERC notified jurisdictional companies that it would "approve applications proposing the recovery of prudently incurred costs necessary to further safeguard the nation's energy systems and infrastructure" in response to the terror attacks of 9/11. FERC also committed to "expedite the processing on a priority basis of any application that would specifically recover such costs from wholesale customers." Companies could propose a surcharge over currently existing rates or some other cost recovery method.[67] In FY2005, the commission processed security cost recovery requests from 14 oil pipelines and 3 natural gas pipelines.[68] FERC's FY2006 annual report stated that "the Commission continues to give the highest priority to deciding any requests made for the recovery of extraordinary expenditures to safeguard the reliability and security of the Nation's energy transportation systems and energy supply infrastructure."[69] FERC's subsequent annual reports do not mention pipeline security.

In February 2003, FERC promulgated a new rule (RM02-4-000) to protect critical energy infrastructure information (CEII). The rule defines CEII as information that "must relate to critical infrastructure, be potentially useful to terrorists, and be exempt from disclosure under the Freedom of Information Act." According to the rule, critical infrastructure is "existing and proposed systems and assets, whether physical or virtual, the incapacity or destruction of which would negatively affect security, economic security, public health or safety, or any combination of those matters." CEII excludes "information that identifies the location of infrastructure." The rule also establishes procedures for the public to request and obtain such critical information, and applies both to proposed and existing infrastructure.[70]

On May 14, 2003, FERC handed down new rules (RM03-4) facilitating the restoration of pipelines after a terrorist attack. The rules allow owners of a damaged pipeline to use blanket certificate authority to immediately start rebuilding, regardless of project cost, even outside existing rights-of-way. Pipeline owners would still need to notify landowners and comply with environmental laws. Prior rules limited blanket authority to $17.5 million projects and 45-day advance notice.[71]

National Transportation Safety Board

The National Transportation Safety Board (NTSB) is an independent federal agency charged with determining the probable cause of transportation accidents (including pipeline accidents), promoting transportation safety, and assisting accident victims and their families. The board's experts investigate significant accidents, develop factual records, and issue safety recommendations to prevent similar accidents from recurring. The NTSB has no statutory authority to regulate transportation, however, so its safety recommendations to industry or government agencies are not mandatory. Nonetheless, because of the board's strong reputation for thoroughness and objectivity, the average acceptance rate for its safety recommendations is over 80%.[72]

San Bruno Pipeline Accident Investigation

In August 2011, the NTSB issued preliminary findings and recommendations from its investigation of the San Bruno Pipeline accident. The investigation included testimony from pipeline company officials, government agency officials (PHMSA, state, and local), as well as testimony from other pipeline experts and stakeholders. In addition to specifics about the San Bruno incident, the hearing addressed more general pipeline issues, including public awareness initiatives, pipeline technology, and oversight of pipeline safety by federal and state regulators.[73] The NTSB's findings were highly critical of the pipeline operator (PG&E) as well as both the state and federal pipeline safety regulators. The board concluded that "the multiple and recurring deficiencies in PG&E operational practices indicate a systemic problem" with respect to its pipeline safety program.[74] The board further concluded that

> the pipeline safety regulator within the state of California, failed to detect the inadequacies in PG&E's integrity management program and that the Pipeline and Hazardous Materials Safety Administration integrity management inspection protocols need improvement. Because the Pipeline and Hazardous Materials Safety Administration has not incorporated the use of effective and meaningful metrics as part of its guidance for performance-based management pipeline safety programs, its oversight of state public utility commissions regulating gas transmission and hazardous liquid pipelines could be improved.

In her opening statement about the San Bruno accident report, NTSB Chairman Hersman summarized the board's findings as "troubling revelations ... about a company that exploited weaknesses in a lax system of oversight and government agencies that placed a blind trust in operators to the detriment of public safety."[75] The NTSB's final accident report "concludes that PHMSA's enforcement program and its monitoring of state oversight programs have been weak and have resulted in the lack of effective Federal oversight and state oversight."[76]

The NTSB issued 39 recommendations stemming from its San Bruno accident investigation, including 20 recommendations to the Secretary of Transportation and PHMSA. These recommendations included:

- Conducting audits to assess the effectiveness of PHMSA's oversight of performance-based pipeline safety programs and state pipeline safety program certification,
- Requiring pipeline operators to provide system-specific information to the emergency response agencies of the communities in which pipelines are located,
- Requiring that automatic shutoff valves or remote control valves be installed in high consequence areas and in class 3 and 4 locations,[77]
- Requiring that all natural gas transmission pipelines constructed before 1970 be subjected to a hydrostatic pressure test that incorporates a spike test,
- Requiring that all natural gas transmission pipelines be configured so as to accommodate internal inspection tools, with priority given to older pipelines, and
- Revising PHMSA's integrity management protocol to incorporate meaningful metrics, set performance goals for pipeline operators, and require operators to regularly assess the effectiveness of their programs using meaningful metrics.[78]

More detailed discussion of the accident findings and the NTSB's recommendations are publicly available in the final accident report.

KEY POLICY ISSUES

The 112th Congress reauthorized the federal pipeline safety program and enacted a number of new pipeline safety provisions. In the context of its continuing oversight of federal pipeline safety and security activities, and in addition to the findings of the NTSB's San Bruno investigation, Congress may focus on certain key issues that have drawn particular attention in recent policy deliberations.

Staffing Resources for Pipeline Safety and Security

The U.S. pipeline safety program employs a combination of federal and state staff to implement and enforce federal pipeline safety regulations. To date, PHMSA has relied heavily on state agencies for pipeline inspections, with only approximately 20% of inspectors in 2011 being federal employees. Some in Congress have criticized inspector staffing at PHMSA as being insufficient to adequately cover pipelines under the agency's jurisdiction, notwithstanding state agency cooperation. In considering potential PHMSA staff increases, three distinct issues are the overall number of federal inspectors, the agency's historical use of staff funding, and the staffing of pipeline safety inspectors among the states.

PHMSA Inspectors

The President's FY2012 budget request listed PHMSA's estimated staffing in 2011 under the continuing resolution as 206 full-time equivalent employees (FTEs). As *Figure 1* shows, the addition of 10 inspection and enforcement staff under P.L. 112-90 would be a modest continuation of staff growth (of mostly inspectors) begun 10 years ago in response to the 1999 Bellingham accident, the terrorist attacks of 9/11, implementation of PHMSA's integrity management regulations, and the continued growth of U.S. pipelines.

Whether an increase of 10 PHMSA pipeline safety staff by 2014, in addition to filling all previously authorized positions, would be the optimal number remains to be seen. The President's FY2012 budget request would fund 290 FTEs in 2013, with most of the staff additions presumably being inspectors.[79] While such an increase would represent a dramatic increase in the number of federal pipeline safety staff available, filling all these positions might pose practical challenges, further discussed below.

PHMSA Staffing Shortfalls

One issue that complicates the PHMSA staffing debate is a long-term pattern of understaffing in the agency's pipeline safety program. At least as far back as 1994, PHMSA's (or RSPA's) actual staffing for pipeline safety as reported in each of its annual budgets requests has fallen short of the level of staffing anticipated in the prior year's budget request. For example, the president's FY2011 budget request for pipeline safety reports 175 actual employees in 2009. However, the FY2010 budget request stated an expectation of 191 employees ("estimated") for 2009.

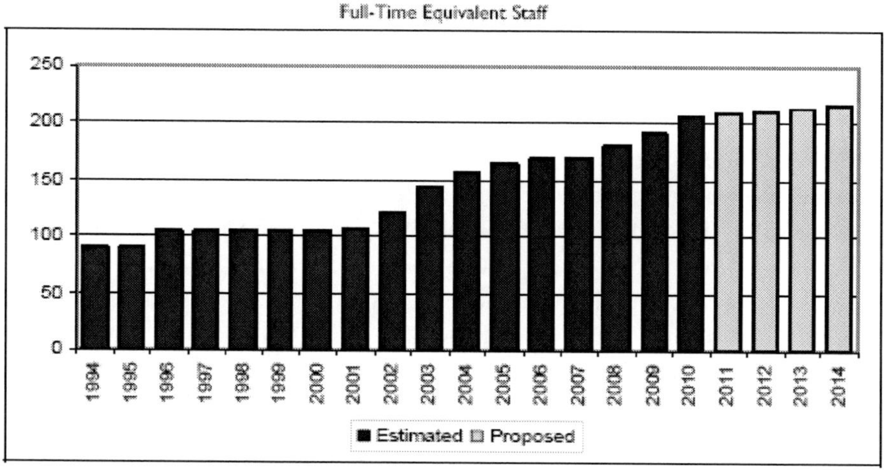

Sources: U.S. Office of Management and Budget, Budget of the United States Government: Appendix, Fiscal Years1996-2011; P.L. 112-90 § 31(b).

Notes: Estimated staff are staff anticipated by the agency as reported in annual budget requests. They may differ from actual staff employed (for the same fiscal year) as reported in subsequent budget requests.

Figure 1. PHMSA Pipeline Safety Staffing, Historical and Proposed under P.L. 112-90.

On this basis, between 2001 and 2009, the agency reported a staffing shortfall averaging approximately 24 employees every year. (Note that, due to this annual shortfall, the FTEs reported in *Figure 1* are higher than the number actually employed by PHMSA.) In testimony before Congress in September 2010, DOT officials reported that PHMSA employed only 110 of 137 inspectors for which it was funded—a shortfall of 27 inspectors.[80] In March 2011, agency officials reported 126 inspectors employed.81 As of February 6, 2012, there were no vacancies posted for a PHMSA pipeline safety inspector position at the USAJobs website.

PHMSA officials offer a number of reasons for the persistent shortfall in inspector staffing. These reasons include a scarcity of qualified inspector job applicants, delays in the federal hiring process during which applicants accept other job offers, and PHMSA inspector turnover— especially to pipeline companies which often hire away PHMSA inspectors for their corporate safety programs. Because PHMSA pipeline inspectors are highly trained by the agency (typically for two years before being allowed to operate independently) they are highly valued by pipeline operators seeking to comply with federal safety regulations. PHMSA officials also cite structural issues associated with the agency's appropriations which can require the use of FTE salary funding to meet other obligations.[82] P.L. 112-90 requires the DOT to report on PHMSA's pipeline staffing shortfalls, including the reasons for such shortfalls, and actions the agency is taking to fill the positions (§ 31(a)).

State Pipeline Inspector Funding

Because state agencies would continue to account for the majority of U.S. pipeline safety inspectors under P.L. 112-90, another important consideration is how the number of state inspectors might be affected by budget shortfalls and possible agency funding cuts faced by many states due to the recent U.S. economic recession. Under P.L. 109-468 (§2(c)), PHMSA is authorized to award grants reimbursing state governments for up to 80% of the cost of the staff, personnel, and activities required to support the federal pipeline safety program (although reimbursement has not reached the 80% level since the passage of the act).

Table 2. Actual vs. Anticipated Pipeline Safety Staff in DOT Budget Requests Full-Time Equivalent Staff

Year	Actual	Anticipated	Difference
1994	62	90	-28
1995	71	90	-19
1996	84	105	-21
1997	92	105	-13
1998	97	105	-8
1999	93	105	-12
2000	97	105	-8
2001	96	107	-11
2002	100	122	-22
2003	111	143	-32
2004	125	156	-31
2005	154	164	-10
2006	139	169	-30
2007	146	170	-24
2008	147	180	-33
2009	175	191	-16

Sources: U.S. Office of Management and Budget, Budget of the United States Government: Appendix, Fiscal Years1996-2011; CRS analysis.

According to DOT, these grant are essential to "enable the states to continue their current programs and hire additional inspectors ... [and] assure that states do not turn over responsibility for distribution pipeline systems to the Federal inspectors," among other reasons.[83] Notwithstanding these federal grants, inspector staffing at state pipeline safety agencies has been negatively affected by state budget deficits. According to the National Association of Pipeline Safety Representatives, pipeline safety employees in 17 states have been furloughed without pay for up to three weeks.[84] PHMSA officials have also reportedly cited unfilled positions among state pipeline safety agencies as eroding the state pipeline safety workforce.[85] Senior DOT officials consider financial problems among state pipeline safety agencies a matter of "great concern" and have granted to states waivers from certain regulatory financial requirements to increase their access to federal grant money.[86] Nonetheless, the future availability of state pipeline safety inspectors remains uncertain. In particular, the possibility that some states may choose to end their roles as agents for the federal pipeline safety program—or that states may lose federal pipeline safety program certification for performance reasons—and thereby shift a greater burden for pipeline inspections back to the federal government, may warrant continued attention from Congress.

TSA Pipelines Security Resources

Similar to its concerns about the adequacy of federal pipeline safety staffing, Congress has long been concerned about staff resources available to implement the nation's pipeline security program. For example, as one Member remarked in 2005, "aviation security has received 90% of TSA's funds and virtually all of its attention. There is simply not enough being done to address ... pipeline security."[87] At a congressional field hearing in April 2010, another Member expressed concern that TSA's pipeline division did not have sufficient staff to carry out a federal pipeline security program on a national scale.[88]

At its current staffing level of 13 FTEs, TSA's Pipelines Security Division has limited field presence. In conducting a pipeline corporate security review, for example, TSA typically sends one to three staff to hold a three- to four-hour interview with the operator's security representatives followed by a visit to only one or two of the operator's pipeline assets.[89] TSA's plan to focus security inspections on the largest pipeline and distribution system operators tries to make the best use of its limited resources. However, there are questions as to whether the agency's CSRs as currently structured allow for rigorous security plan verification and a credible threat of enforcement. The limited number of CSRs the agency can complete in a year is a particular concern. According to a 2009 GAO report, "TSA's pipeline division stated that they would like more staff in order to conduct its corporate security reviews more frequently," and "analyzing secondary or indirect consequences of a terrorist attack and developing strategic risk objectives required much time and effort."[90] P.L. 110-53 specifically authorized funding of $2 million annually through FY2010 for TSA's pipeline security inspections and enforcement program (§1557(e)). It is an open question whether $2 million annually, were this funding to continue, would be sufficient to enable TSA to meet congressional expectations for federal pipeline security activities.

Since both PHMSA and TSA have played important roles in the federal pipeline security program, with TSA the designated lead agency since 2002, Congress has raised questions about the appropriate responsibilities and division of pipeline security authority between them.[91] According to TSA, the two agencies have established "a 24/7 communication and coordination relationship in regards to all pipeline security and safety incidents."[92] Nonetheless, given the limited staff in TSA's pipeline security division, and the comparatively large pipeline safety staff (especially inspectors) in PHMSA, legislators have considered whether the TSA-PHMSA pipeline security relationship optimally aligns staff resources across both agencies to fulfill the nation's overall pipeline safety and security mission.[93] The Transportation Security Administration Authorization Act of 2011 (H.R. 3011) would mandate a study regarding the relative roles and responsibilities of the Department of Homeland Security and the Department of Transportation with respect to pipeline security (§325).[94]

Automatic Shutoff Valves for Transmission Pipelines

In the 2010 San Bruno pipeline accident, natural gas continued to flow from the pipeline for nearly two hours after the initial explosion—fueling the intense fire, hindering emergency response, and increasing fire damage. The long duration of flowing gas reportedly was due to delays in the closing of manually operated valves by the pipeline operator, and may have been exacerbated by inadequate employee training in valve closure procedures.95 Consequently, some advocates have called for widespread installation of remotely or automatically controlled valves in natural gas and hazardous liquids transmission pipelines. As noted earlier, the NTSB has recommended the installation of such valves in all "high consequence" and relatively more populated areas. P.L. 112-90 requires automatic or remote-controlled shut-off valves, or equivalent technology, where economically, technically, and operationally feasible on transmission pipeline facilities constructed or entirely replaced (§4(1)). The act also requires a study on the ability of transmission pipeline operators to respond to a hazardous liquid or natural gas release from pipelines located in high consequence areas (§4(2)).

Previous Consideration

The possibility of requiring remotely controlled or automatic shut off valves for natural gas pipelines is not new. Congress previously considered such requirements in reaction to a 1994 natural gas pipeline fire in Edison, NJ, similar to the San Bruno accident in which it took the pipeline operator 21/2 hours to close its manually operated valves.[96] In 1995, during the 104th Congress, H.R. 432 and S. 162 would have required the installation of remotely or automatically controlled valves in natural gas pipelines "wherever technically and economically feasible" (§11). Under the Accountable Pipeline Safety and Partnership Act of 1996 (P.L. 104-304), Congress mandated a DOT assessment of remotely controlled valves (RCVs) on interstate natural gas pipelines, and empowered the agency to require such valves if appropriate based upon its findings (§4(h)).

The DOT's assessment, released in 1999, reported that installation of RCVs would provide only "a small benefit from reduced casualties because virtually all casualties from a rupture occur before an RVC could be activated."[97] Moreover, the DOT reported that it lacked data to compare pipeline fire property damage with and without RCVs. Nonetheless, the DOT study advocated the deployment of RCVs, at least in some gas pipeline locations.

> We have found that RCVs are effective and technically feasible, and can reduce risk, but are not economically feasible. We have also found that there may be a public perception that RCVs will improve safety and reduce the risk from a ruptured gas pipeline. We believe there is a role for RCVs in reducing the risk from certain ruptured pipelines and thereby minimizing the consequences of certain gas pipeline ruptures.... Any fire would be of greater intensity and would have greater potential for damaging surrounding infrastructure if it is constantly replenished with gas. The degree of disruption in heavily populated and commercial areas would be in direct proportion to the duration of the fire. Although we lack data enabling us to quantify these potential consequences, we believe

> them to be significant nonetheless, and we believe RCVs may provide the best means for addressing them.[98]

Notwithstanding this conclusion, the DOT has not mandated the use of RCVs in natural gas transmission pipelines.

The natural gas pipeline industry historically has objected to federal mandates to install remotely controlled or automated valves. Although pipeline operators already employ such valves under specific circumstances, such as in hard-to-access locations or at compressor stations, they have opposed the installation of such valves more widely throughout their pipeline systems on the grounds that they are usually not cost-effective. They also argue that such valves do not always function properly, would not prevent natural gas pipeline explosions (which cause most fatalities), and are susceptible to false alarms, needlessly shutting down pipelines and disrupting critical fuel supplies.[99] Automatic valves, in particular, may be susceptible to unnecessary closure, potentially disrupting critical flows of natural gas to distribution utilities and—as a result—increasing safety risks associated with residential furnace relighting, among other concerns.[100] Some operators also claim higher maintenance costs for valves that are not manually operated.

Remotely Controlled Valves for Liquids Pipelines

The use of remotely controlled or automatic valves has also been a long-standing consideration for hazardous liquid pipeline systems. The National Transportation Safety Board (NTSB) began to address the need for rapid shutdown of failed hazardous liquid pipelines using remotely controlled or automatic valves in the 1970s.[101] In 1987, the NTSB recommended that the DOT "require the installation of remote-operated valves on pipelines that transport hazardous liquids, and base the spacing of remote-operated valves on the population at risk."[102] The Pipeline Safety Act of 1992 (P.L. 102-508) required the DOT to assess the effectiveness of "emergency flow restricting devices (including remotely controlled valves and check valves)" on hazardous liquid pipelines, and required the DOT to "issue regulations prescribing the circumstances under which operators of hazardous liquid pipeline facilities must use emergency flow restricting devices" (§212). Notwithstanding this congressional mandate, the NTSB found the DOT's efforts to promote the use of such devices inadequate. In 1996, the NTSB stated that the DOT "has performed studies, conducted research, and sought industry input, but has failed to carry through and develop requirements for leak detection and rapid shutdown of failed pipelines."[103] In its integrity management regulations, issued in December 2000, the DOT opted to leave the decision whether to install emergency flow restricting devices up to pipeline operators.[104]

Valve Replacement Costs

Cost would be a major factor in a broad national program to retrofit manual valves with remotely controlled or automatic valves. For example, in the interstate natural gas pipeline network, valves are typically installed every 5 to 20 miles. Assuming a 10-mile separation between valves, the nation's 306,000 mile gas transmission system contains over 30,000 valves. The spacing of valves can be much closer together in particular pipeline systems, however, such as systems located in more populated areas. In October 2010 PG&E reported 300 valves that could be candidates for automation in approximately 565 miles of high-consequence area pipelines in its California system.[105]

The potential costs of retrofitting manual valves vary greatly by pipeline and specific location. A 1998 Southwest Research Institute report estimated a cost of $32,000 (approximately $40,000 in 2010 dollars) per valve for retrofitting 30-inch pipeline valves to make them remotely controlled.[106] The DOT's 1999 study reported an average cost of $83,000 (approximately $100,000 in 2010 dollars) for Texas Eastern Transmission Corporation (TETCO) to retrofit 90 existing valves in a large part of its pipeline system.[107] PG&E estimates the average cost of retrofitting an automatic or remotely controlled valve on an existing large diameter pipeline at approximately $750,000, but ranging as low as $100,000 and as high as $1.5 million.[108]

Applying, for illustration, a $100,000 cost to some 30,000 valves yields $3.0 billion in capital investment required, not counting any higher future maintenance expenses. The American Gas Association reportedly has estimated the cost of replacing manual valves with automatic valves nationwide at $12 billion.[109] Even if such valve retrofits were required only in heavily populated areas, industry costs could still be hundreds of millions of dollars—a significant cost to the pipeline industry and therefore likely to increase rates for pipeline transportation of natural gas. To the extent that some pipeline systems, like PG&E's, contain more valves then others per mile of pipe, they could be disproportionately affected. Gas pipeline service interruptions would also be an issue as specific lines could be repeatedly taken out of service during the valve retrofit process. The hazardous liquids pipeline industry could face capital costs and service interruptions of the same magnitude if required to do a widespread valve retrofit on existing lines.

Additional right-of-way costs, environmental impacts, and construction accidents associated with the valve replacements could also be a consideration. For new pipelines, the incremental costs of installing remotely controlled or automatic valves instead of manual valves would be lower than in the retrofit case, but could still increase future pipeline costs.

SCADA and Leak Detection System Requirements

To effectively reduce the impact of pipeline accidents, installing remotely controlled or automatic valves may require associated investments in supervisory control and data acquisition (SCADA) systems along with other operational changes to improve leak detection. As one pipeline expert has stated,

> The pipeline operator's focus on keeping the pipeline system operating and the lack of remotely-operable valves are the primary factors that control the quantity of product released after a rupture or leak. Even with remote control valves this relationship will not change unless the pipeline is equipped with a reliable leak detection subsystem that works with the SCADA system and [unless] those who control pipeline operations are trained for and dedicated to minimizing product release (safety and environmental mindset) rather than trained for and dedicated to keeping the system operating (economic mindset).[110]

In its report about a 1996 pipeline accident in Tiger Pass, LA, the NTSB similarly concluded that the operator's "delay in recognition ... that it had experienced a pipeline rupture at Tiger Pass was due to the piping system's dynamics during the rupture and to the design of the company's SCADA system."[111] Estimates of converting manual valves may, therefore, need to account for the costs of SCADA changes, leak detection systems, and associated training. These costs may also include significant reliability and security components, since increasing reliance upon new or expanded SCADA systems may also expose pipeline systems to greater risk from operating software failure or cyberterrorism.[112]

Consistent with the concerns above, P.L. 112-90 requires a DOT analysis of the technical limitations of leak detection systems as well as the feasibility of establishing standards for such systems (§8(a)). After congressional review of this analysis, the act authorizes the DOT to issue new leak detection standards if "practicable" (§8(b)).

Public Perceptions

Some stakeholders have argued that public perceptions of improved pipeline safety and control are the highest perceived benefit of remotely controlled or automatic valves.[113] Although the value of these perceptions is hard to quantify (and, therefore, not typically reflected in cost-effectiveness studies), the importance of public perception and community acceptance of pipeline infrastructure can be a significant consideration in pipeline design, expansion, and regulation. In 2001, a representative of the National Association of Regulatory Utility Commissioners testified before Congress that "the main impediment to siting energy infrastructure is the great difficulty getting public acceptance for needed facilities."[114] Likewise, the National Commission on Energy Policy stated in its 2006 report that energy-facility siting is "a major cross-cutting challenge for U.S. energy policy," largely because of public opposition to new energy projects and other major infrastructure.[115]

One result of public concern about pipeline safety has been to prevent new pipeline siting in certain localities, and to increase pipeline development time and costs in others. In a 2006 report, for example, the Energy Information Administration (EIA) stated that "several major projects in the Northeast, although approved by FERC, have been held up because of public opposition or non-FERC regulatory interventions."[116] In the specific case of the Millennium Pipeline, proposed in 1997 to transport Canadian natural gas to metropolitan New York, developers did not receive final construction approval for nine years, largely because of community resistance to the pipeline route.[117] Numerous other proposed pipelines, especially in populated areas, have faced similar public acceptance barriers.[118] Controversy surrounding the proposed Keystone XL pipeline project, discussed below, is only the most recent example of how the development of major pipeline projects may be influenced by public opinion. Even where there is federal siting authority, as is the case for interstate natural gas pipelines, community stakeholders retain many statutory and regulatory avenues to affect energy infrastructure decisions. Consequently, the public perception value of remotely controlled or automatic pipeline valves may need to be accounted for, especially with respect to its implications for general pipeline development and operations.

Natural Gas Distribution Excess Flow Valves

While the San Bruno, CA, and Edison, NJ, gas pipeline accidents focused attention on automatic valves in large diameter transmission pipelines, this technology also applies to smaller gas distribution lines serving individual buildings. In natural gas distribution systems, "excess flow" valves are safety devices which can automatically shut off pipeline flow in the event of a leak. In this way, the valves can minimize the release of natural gas during a pipeline accident, thereby reducing the likelihood or severity of a fire or explosion. PHMSA issued new standards requiring the installation of excess flow valves on new gas distribution lines in single-family homes as part of its final rule for natural gas distribution integrity management programs on December 3, 2009.[119] P.L. 112-90 authorizes regulation, "if appropriate," requiring excess flow valves for new or entirely replaced distribution branch pipelines, as well as for service lines to multi-family residential buildings and small commercial facilities. Although smaller in scale, automatic valves in distribution lines raise the same cost and safety tradeoffs as automatic valves in large diameter pipelines (§22).

PHMSA Penalties and Pipeline Safety Enforcement

The adequacy of the PHMSA's enforcement strategy has been an ongoing focus of congressional oversight.[120] Provisions in the Pipeline Safety Improvement Act of 2002 (P.L. 107-355) put added scrutiny on the effectiveness of the agency's enforcement strategy and assessment of civil penalties (§8). In April 2006, PHMSA officials testified before Congress that the agency had institutionalized a "tough-but-fair" approach to enforcement, "imposing and collecting larger penalties, while guiding pipeline operators to enhance higher performance."[121] According to the agency, $4.6 million in proposed civil penalties in 2005 was three times greater than penalties proposed in 2003, the first year higher penalties could be imposed under P.L. 107-355 (§8(a)).[122] Proposed penalties totaled $4.5 million in 2010.[123] Proposed penalties in 2011 totaled $3.7 million, with an average penalty of approximately $65,500.[124] P.L. 112-90 increases the maximum civil penalty from $1.0 million to $2.0 million for a related series of major consequence violations, such as those causing serious injuries, deaths, or environmental harm (§2(a)).

Although PHMSA's imposition of pipeline safety penalties increased quickly after P.L. 107-355 was enacted, the role of federal penalties in promoting greater operator compliance with pipeline safety regulations is not always clear. To understand the potential influence of penalties on operators, it can be helpful to put PHMSA fines in the context of the overall costs to operators of a pipeline release. Pipeline companies, seeking to generate financial returns for their owners, are motivated to operate their pipelines safely (and securely) for a range of financial reasons. While these financial considerations certainly include possible PHMSA penalties, the costs of a pipeline accident may also include fines for violations of environmental laws (federal and state), the costs of spill response and remediation, penalties from civil litigation, the value of lost product, costs for pipeline repairs and modifications (e.g., to resolve federal regulatory interventions), and other costs. Depending upon the severity of a pipeline release, these other costs may far exceed pipeline safety fines, as illustrated by the following examples.

- *Kinder Morgan.* In April 2006 Kinder Morgan Energy Partners entered into a consent agreement with PHMSA to resolve a corrective action order stemming from three hazardous liquid spills in 2004 and 2005 from the company's Pacific Operations pipeline unit.[125] According to the company, the agreement would require Kinder Morgan to spend approximately $26 million on additional integrity management activities, among other requirements.[126] Under a 2007 settlement agreement with the U.S. Justice Department and the state of California, Kinder Morgan also agreed to pay approximately $3.8 million in civil penalties for violations of environmental laws and approximately $1.5 million related to response and remediation associated with these spills. The spills collectively released approximately 200,000 gallons of diesel fuel, jet fuel, and gasoline.[127] This volume of fuel would have a product value on the order of $0.5 million based on typical wholesale market prices at the time of the spills.
- *Plains All American.* In 2010, Plains All American Pipeline agreed to spend approximately $41 million to upgrade 10,420 miles of U.S. oil pipeline to resolve Clean Water Act (CWA) violations for 10 crude oil spills in Texas, Louisiana, Oklahoma, and Kansas from 2004 through 2007. Among these upgrades, the company agreed to spend at least $6 million on equipment and materials for internal corrosion control and surveys on at least 2,400 miles of pipeline. The company was required to pay $3.25 million civil penalty associated with the CWA violations.[128]
- *Enbridge.* Enbridge Energy Partners estimated expenses of $475 million to clean up two oil spills on its Lakehead pipeline system in 2010, including the spill in Marshall, MI. This estimate did not include fines or penalties which might also be imposed in connection with the spills. The pipeline operator also reported $16 million in lost

revenue from pipeline shipments it could not redirect to other lines while the Lakehead system was out of service.[129] The full impact of these expenditures on the company's business is unclear, however, as Enbridge stated in a subsequent quarterly report that "substantially all of the costs" related to its 2010 oil pipeline spills "will ultimately be recoverable under our existing insurance policies."[130]

- *Olympic Pipe Line.* After the 1999 Bellingham pipeline accident, Olympic Pipe Line Company and associated defendants reportedly agreed to pay a $75 million settlement to the families of two children killed in the accident.[131]
- *El Paso.* In 2002, El Paso Corporation settled wrongful death and personal injury lawsuits stemming from the 2000 natural gas pipeline explosion near Carlsbad, NM, which killed 12 campers.[132] Although the terms of those settlements were not disclosed, two additional lawsuits sought a total of $171 million in damages.[133] However, El Paso's June 2003 quarterly financial report stated that "our costs and legal exposure ... will be fully covered by insurance."[134]

PHMSA Penalties in Perspective

The threat of safety enforcement penalties is often considered one of the primary tools available to pipeline safety regulators to ensure operator compliance with safety requirements. However, as the examples above suggest, pipeline safety fines, even if they were raised to $2.0 million for major violations, could still account for only a limited share of the financial impact of future pipeline releases. So, it is not clear how large an effect increasing PHMSA's authorized fines, alone, might have on operator compliance.

On the other hand, the authority of PHMSA to influence pipeline operations directly—for example, through corrective action orders or shutdown orders in the event of a pipeline failure—can have a large financial impact on a pipeline operator in terms of capital expenditures or lost revenues. Indeed, some have suggested that this operational authority is the most influential component of PHMSA's pipeline safety enforcement strategy. Therefore, as Congress continues its oversight of PHMSA's enforcement activities, and as it considers new proposals to increase compliance with federal pipeline safety regulations, Congress may evaluate how PHMSA's authorities to set standards, assess penalties, and directly affect pipeline operations may reinforce one another to improve U.S. pipeline safety.

Regulation of Canadian Oil/Tar Sands Crude Pipelines

Canadian oil exports to the United States have been increasing rapidly, primarily due to growing output from the oil sands in Western Canada.[135] Oil sands (also referred to as tar sands) are a mixture of clay, sand, water, and heavy black viscous oil known as bitumen. Oil sands are processed to extract the bitumen, which can then be upgraded into a product that is suitable for pipeline transport. Canada's oil sands production can be exported as either a light, upgraded synthetic crude ("syncrude") or a heavy crude oil that is a blend of bitumen diluted with lighter hydrocarbons ("dilbit") to ease transport. The bulk of oil sands' supply growth is expected to be in the form of the latter.[136] Five major pipelines have been constructed in recent years to link the oil sands region to markets in the United States. A sixth pipeline, Keystone XL, was rejected in January 2012 by the U.S. State Department, although the developer plans to reapply for a federal permit with a modified route.[137] If ultimately approved and constructed, Keystone XL would bring Canada's total U.S. petroleum export capacity to over 4.1Mbpd, enough capacity to carry over a third of current U.S. petroleum imports.[138]

This expansion of petroleum pipelines from Canada has generated considerable controversy in the United States. One specific area of concern has been perceived new risks to pipeline integrity of transporting heavy Canadian crudes. Some opponents of the new Canadian oil pipelines, notably the Natural Resources Defense Council (NRDC), has argued that these pipelines could be more likely to fail and cause environmental damage than other crude oil pipelines because the bitumen mixtures they would carry are "significantly more corrosive to pipeline systems than conventional crude," among other reasons.[139] The council has called for a moratorium on approving new oil pipelines from oil sands regions, and a review of existing pipeline permits, until these safety concerns are researched further and addressed in federal environmental and safety studies. Canadian officials and other stakeholders have rejected these arguments, however, citing factual inaccuracies and a flawed methodology in the NRDC analysis, which compares pipeline spill rates in Canada to those in the United States.[140]

Some in Congress have called for a review of PHMSA regulations to determine whether new regulations for Canadian heavy crudes are needed to account for any unique properties they may have. Accordingly, P.L. 112-90 requires PHMSA to review whether current regulations are sufficient to regulate pipelines transmitting "diluted bitumen," and analyze whether such oil presents an increased risk of release (§16).

Pipeline Security Regulations

As noted earlier in this report, federal pipeline security activities to date have relied upon voluntary industry compliance with PHMSA security guidance and TSA security best practices. By initiating this voluntary approach, PHMSA sought to speed adoption of security measures by industry and avoid the publication of sensitive security information (e.g., critical asset lists) that would normally be required in public rulemaking.[141] Provisions in P.L. 109-468 require the DOT Inspector General to "address the adequacy of security standards for gas and oil pipelines" (§23(b)(4)). P.L. 110-53 similarly directs TSA to promulgate pipeline security regulations and carry out necessary inspection and enforcement—if the agency determines that regulations are appropriate (§1557(d)). Addressing this issue, the 2008 IG report states that

> TSA's current security guidance is not mandatory and remains unenforceable unless a regulation is issued to require industry compliance.... PHMSA and TSA will need to conduct covert tests of pipeline systems' vulnerabilities to assess the current guidance as well as the operators' compliance.[142]

Although TSA's FY2005 budget justification stated that the agency would "issue regulations where appropriate to improve the security of the [non-aviation transportation] modes," the agency has not done so for pipelines, and it is not currently working on such regulations.[143] The pipeline industry has expressed concern that new security regulations and related requirements may be "redundant" and "may not be necessary to increase pipeline security."[144] The PHMSA Administrator, in 2007, testified that enhancing security "does not necessarily mean that we must impose regulatory requirements."[145] TSA officials have questioned the IG assertions regarding pipeline security regulations, particularly the IG's call for covert testing of pipeline operator security measures. They have argued that the agency is complying with the letter of P.L. 110-53 and that its pipeline operator security reviews are more than paper reviews.[146] In accordance with P.L. 110-53 (§1557 (b)), the TSA has been implementing a multi-year program of pipeline system inspections, including documentation of findings and follow up reviews.[147] In its oversight of potential pipeline security regulations, Congress may evaluate the effectiveness of the current voluntary pipeline security standards based on findings from the TSA's CSR reviews, pipeline inspections, and any future DOT Inspector General reports.

Additional Issues

In addition to the issues mentioned above, Congress may consider several issues related to proposed legislation or otherwise raised by pipeline stakeholders.

Accuracy and Completeness of Pipeline System Records

On January 3, 2011, as a response to its initial investigation of the San Bruno pipeline accident, the NTSB issued urgent new safety recommendations "to address record-keeping problems that could create conditions in which a pipeline is operated at a higher pressure than the pipe was built to withstand."[148] The NTSB issued these recommendations after it had concluded that there were significant errors in the records characterizing the San Bruno pipeline, and that "other pipeline operators may have discrepancies in their records that could potentially compromise the safe operation of pipelines throughout the United States."[149] PHMSA officials have also testified that some operators may not be collecting all the pipeline system data necessary to fully evaluate safety and compliance with federal regulations.[150] In 2006, questions were raised about the accuracy of pipeline location data provided by operators and maintained by PHMSA in the National Pipeline Mapping System.[151] At the time, agency officials reportedly acknowledged limitations in NPMS accuracy, but did not publicly discuss plans to address them. P.L. 112-90 authorizes PHMSA to collect additional geospatial and technical data from pipeline operators to achieve the purposes of the NPMS (§11(a) and §12). Congress may review whether these or other statutory measures are sufficient to verify that pipeline operator information is complete and correct, particularly for older parts of the pipeline network.

Mandatory Internal Inspection or Hydrostatic Testing

Some proposals would increase requirements for pipeline operators to conduct internal inspections of transmission pipelines using "smart pigs," robotic devices sent through pipelines to take physical measurements continuously along the way.[152] In its San Bruno accident investigation report, the NTSB has recommended that all natural gas transmission pipelines be configured to accommodate such internal inspection tools. However, experts note that there are different pipeline inspection techniques with overlapping capabilities and different strengths.[153] While an effective technology for detecting corrosion in many applications, smart pigs have limitations as a general tool for assessing the integrity of pipelines. For example, although smart pigs may be good corrosion detectors, they are still a developing technology and may be somewhat less effective in detecting other types of pipeline anomalies (e.g., cracks). Operators also maintain that smart pigging may be less useful for predicting future problems with pipeline integrity than other federally approved maintenance techniques like “direct assessment” (49 C.F.R. 192.903) wherein pipelines are examined externally based on risk data and other factors.[154] Furthermore, because many older pipelines contain sharp turns and other obstructions due to historical construction techniques, they cannot accommodate smart pig devices without significant and costly pipeline modifications to make them more “piggable.” Consequently, some industry stakeholders caution against unrealistic expectations for the capabilities of smart pigs as a stand-alone pipeline inspection tool.[155]

As an alternative to internal inspection where such inspection cannot currently be performed, some policy makers have called for mandatory hydrostatic testing of pipelines to verify their integrity. Hydrostatic testing involves filling a pipeline with water under pressure greater than the anticipated operating pressure to determine if it is structurally sound and does not leak. Such testing is common for new pipelines that have not yet entered service. Because it uses only water, hydrostatic testing poses relatively little direct risk to the public or the environment, but when used for operating pipelines it necessarily interrupts pipeline service. Injecting water into pipelines is also costly, and may create safety problems since water is corrosive and may be difficult to remove completely from a pipeline once testing is completed.[156] Nonetheless, as noted above, the NTSB has recommended that all natural gas transmission pipelines constructed before 1970 be subjected to hydrostatic pressure tests. P.L. 112-90 requires verification of maximum allowable operating pressure for all natural gas transmission pipelines “as expeditiously as economically feasible” (§23(a)). The act also authorizes regulations for pressure verification that “shall consider ... pressure testing; and ... other alternative methods, including in-line inspections” (§23(a)). As Congress examines any new federal requirements for pipeline inspection, it may consider smart pig devices and hydrostatic testing as only two techniques in a portfolio of maintenance practices operators may need to employ to ensure their pipelines are physically sound.

Emergency Response Plan Disclosure

Federal regulations require pipeline operators to prepare emergency response plans for pipeline spills and to make those plans available for inspection by PHMSA and local emergency response agencies (49 C.F.R. 192.605). Some stakeholders have proposed that these plans also be made available to the public to allow for additional review of their adequacy and to provide better risk and response information to people living near pipelines.[157] Operators reportedly have resisted such disclosures on the grounds that their emergency response plans contain confidential customer and employee information.[158] They also raise concerns that the plans contain security-sensitive information about pipeline vulnerabilities and spill scenarios which could be useful to terrorists.[159] P.L. 112-90 requires PHMSA to collect and maintain copies of pipeline emergency plans for public availability excluding any proprietary or security-sensitive information (§6(a)). As oversight of this issue continues, Congress may consider the tradeoffs between public awareness and pipeline security in a general operating environment where both safety and security hazards may be significant.

Pipeline Water Crossings

The 2011 oil spill into the Yellowstone River near Laurel, MT, appears to have been the result of the buried oil pipeline becoming exposed due to scouring of the river bottom during unusually heavy flooding.[160] Prior to the flooding, a depth-of-cover survey by the operator verified that the pipeline was at least five feet below the riverbed, exceeding a four-foot minimum cover requirement in PHMSA regulations.[161] Because the four-foot requirement appears to have been insufficient to prevent riverbed pipeline exposure in this case, some policy makers have called for a review of pipeline river crossings and associated safety requirements nationwide. P.L. 112-90 mandates a review of the adequacy of PHMSA regulations with respect to pipelines that cross inland bodies of water at least 100 feet wide and, based on the review's findings, requires PHMSA to develop legislative recommendations for changing existing regulations (§28(a)).

CONCLUSION

Both government and industry have taken numerous steps to improve pipeline safety and security over the last 10 years. While stakeholders across the board agree that federal pipeline safety programs have been on the right track, major pipeline incidents in 2010 and 2011 suggest there continues to be significant room for improvement. Likewise the threat of terrorist attack on U.S. pipeline infrastructure remains a concern.

As Congress oversees the federal pipeline safety program and the federal role in pipeline security, key issues of focus may be pipeline agency staff resources, automatic pipeline shutoff valves, penalties for safety violations, safety regulations for oil sands crudes, and the possible need for pipeline security regulations, among other concerns. In addition to these specific issues, Congress may assess how the various elements of U.S. pipeline safety and security activity fit together in the nation's overall strategy to protect transportation infrastructure. Pipeline safety and security necessarily involve many groups: federal agencies, oil and gas pipeline associations, large and small pipeline operators, and local communities. Reviewing how these groups work together to achieve common goals could be an oversight challenge for Congress.

End Notes

[1] Hazardous liquids primarily include crude oil, gasoline, jet fuel, diesel fuel, home heating oil, propane, and butane. Other hazardous liquids transported by pipeline include anhydrous ammonia, carbon dioxide, kerosene, liquefied ethylene, and some petrochemical feedstocks.

[2] Pipeline and Hazardous Materials Safety Administration, "Natural Gas Transmission, Gas Distribution, and Hazardous Liquid Pipeline Annual Mileage," online table, September 21, 2010, http://www.phmsa.dot.gov/pipeline/ library/data-stats.

[3] Richard A. Rabinow, "The Liquid Pipeline Industry in the United States: Where It's Been, Where It's Going," Prepared for the Association of Oil Pipe Lines, April 2004, p. 4.

[4] Energy Information Administration, "Estimated Natural Gas Pipeline Mileage in the Lower 48 States, Close of 2008," online table, 2011, http://www.eia.doe.gov/pub/oil_gas/natural_gas/ analysis_publications/ngpipeline/mileage.html.

[5] Pipeline and Hazardous Materials Safety Administration, "Natural Gas Transmission, Gas Distribution, and Hazardous Liquid Pipeline Annual Mileage," Web table, March 15, 2011, http://www.phmsa.dot.gov/portal/site/ PHMSA/menuitem.7c371785a639f2e55cf2031050248a0c/?vgnextoid= 3b6c03347e4d8210VgnVCM1000001ecb7898RCRD&vgnextchannel= 3b6c03347e4d8210VgnVCM1000001ecb7898RCRD&vgnextfmt=print.

[6] Energy Information Administration, "About U.S. Natural Gas Pipelines," June 2007, pp. 1, 29, http://www.eia.doe.gov/pub/oil_gas/natural_gas/analysis_publications/ngpipeline/fullversion.pdf.

[7] Pipeline and Hazardous Materials Safety Administration, September 21, 2010, http://www.phmsa.dot.gov/pipeline/ library/data-stats.

[8] There are also approximately 6,300 miles of offshore gathering pipelines. Gathering pipelines in the Outer Continental Shelf regulated by the Department of the Interior are outside the scope of this report.

[9] Pipeline and Hazardous Materials Safety Administration, "Significant Pipeline Incidents," web page, February 2, 2012, http://primis.phmsa.dot.gov/comm/reports/safety/SigPSI.html.

[10] Ibid.

[11] J.L. Shreeve, "Science & Technology: The Enemy Within," The Independent. London, UK, May 31, 2006, p. 8.

[12] Government Accountability Office (GAO), Security Assistance: Efforts to Secure Colombia's Caño Limón-Coveñas Oil Pipeline Have Reduced Attacks, but Challenges Remain, GAO-05-971, September 2005, p. 15; Stratfor Forecasting, Inc.," Colombia: The FARC's Low-Level Pipeline Campaign," Stratfor Today, June 23, 2008. http://www.stratfor.com/analysis/colombia_farcs_low_level_pipeline_campaign?ip_auth_redirect=1.

[13] President's Commission on Critical Infrastructure Protection, Critical Foundations: Protecting America's Infrastructures, Washington, DC, October 1997.

[14] Katehrine Houreld, "Militants Say 3 Nigeria Pipelines Bombed," Associated Press, May 8, 2007.

[15] Reed Johnson, "Six Pipelines Blown Up in Mexico," Los Angeles Times, September 11, 2007. p A-3.

[16] U.S. Department of Justice, "Four Individuals Charged in Plot to bomb John F. Kennedy International Airport," Press release, June 2, 2007.

[17] Ben Gelinas, "New Letter Threatens Resumption of 'Action' against B.C. Pipelines," Calgary Herald, April 15, 2010.

[18] Steve Fainaru and William Booth, "Mexico's Drug Cartels Siphon Liquid Gold," Washington Post, December 13, 2009.

[19] "Already Hard at Work on Security, Pipelines Told of Terrorist Threat," Inside FERC, McGraw-Hill Companies, January 3, 2002.

[20] Alyeska Pipeline Service Co., Internet page, Anchorage, AK, February 8, 2009, http://www.alyeska-pipe.com/ about.html.

[21] David S. Cloud, "A Former Green Beret's Plot to Make Millions Through Terrorism," Ottawa Citizen, December 24, 1999, p. E15.

[22] Y. Rosen, "Alaska Critics Take Potshots at Line Security," Houston Chronicle, February 17, 2002.

[23] Wesley Loy, "Web Post Urges Jihadists to Attack Alaska Pipeline," Anchorage Daily News, January 19, 2006.

[24] U.S. Attorney's Office, Middle District of Pennsylvania, "Man Convicted of Attempting to Provide Material Support to Al-Qaeda Sentenced to 30 Years' Imprisonment," Press release, November 6, 2007; A. Lubrano and J. Shiffman, "Pa. Man Accused of Terrorist Plot," Philadelphia Inquirer, February 12, 2006, p. A1.

[25] Transportation Security Administration, Office of Intelligence, Pipeline Threat Assessment, January 18, 2011, p. 3.

[26] Carol Cratty, "Man Accused in Attempted Bombing of Oklahoma Gas Pipeline," CNN, August 12, 2011.

[27] Safety and security of liquefied natural gas (LNG) facilities used in gas pipeline transportation is regulated under CFR Title 49, Part 193.

[28] U.S. Office of Management and Budget, Budget of the United States Government, Fiscal Year 2011: Appendix, February 2010, p. 989.

[29] John D. Porcari, Deputy Secretary, U.S. Department of Transportation, Testimony before the House Committee on Transportation and Infrastructure, Hearing on the Enbridge Pipeline Oil Spill in Marshall, MI, September 15, 2010.

[30] 49 U.S.C. 601. States may recover up to 50% of their costs for these programs from the federal government.

[31] U.S. Office of Management and Budget, Budget of the United States Government, Fiscal Year 2012: Appendix February 2012, p. 1045.

[32] Pipeline and Hazardous Material Safety Administration (PHMSA), "PHMSA Pipeline Safety Program: Summary of Enforcement Actions," Web page, January 9, 2012, http://primis.phmsa.dot.gov/comm/reports/enforce/ Actions_opid_0.html?nocache=8828.

[33] Pipeline and Hazardous Material Safety Administration (PHMSA), "PHMSA Pipeline Safety Program: Summary of Cases Involving Civil Penalties," Web page. January 9, 2012, http://primis.phmsa.dot.gov/comm/reports/enforce/ CivilPenalty_opid_0.html?nocache=9288#_TP_1_tab_2.

[34] Presidential Decision Directive 63, Protecting the Nation's Critical Infrastructures, May 22, 1998.

[35] In November 2004, the President signed the Norman Y. Mineta Research and Special Programs Improvement Act (P.L. 108-426), which eliminated RSPA and placed the Office of Pipeline Safety under the new Pipeline and Hazardous Material Safety Administration. This restructuring did not significantly alter the authorities or activities of the OPS.

[36] Research and Special Programs Administration (RSPA), RSPA Pipeline Security Preparedness, December 2001.

[37] Ellen Engleman, Administrator, Research and Special Programs Administration (RSPA), statement before the Subcommittee on Energy and Air Quality, House Energy and Commerce Committee, March 19, 2002.

[38] Ellen Engleman, Administrator, Research and Special Programs Administration (RSPA), statement before the Subcommittee on Highways and Transit, House Transportation and Infrastructure Committee, February 13, 2002.

[39] James K. O'Steen, Research and Special Programs Administration (RSPA), Implementation of RSPA Security Guidance, presentation to the National Association of Regulatory Utility Commissioners, February 25, 2003.

[40] Office of Pipeline Safety (OPS), personal communication, June 10, 2003.

[41] P.L. 107-355 encourages the implementation of state "one-call" excavation notification programs (§ 2) and allows states to enforce "one-call" program requirements. The act expands criminal responsibility for pipeline damage to cases where damage was not caused "knowingly and willfully" (§ 3). The act adds provisions for ending federal-state pipeline oversight partnerships if states do not comply with federal requirements (§ 4).

[42] A 2006 Government Accountability Office (GAO) report found that PHMSA's gas integrity management program benefitted public safety, although the report recommended revisions to PHMSA's performance measures. See GAO, "Natural Gas Pipeline Safety: Integrity Management Benefits Public Safety, but Consistency of Performance Measures Should Be Improved," GAO-06-946, September 8, 2006, pp. 2-3.

[43] PHMSA issued final regulations requiring operators of natural gas distribution pipelines to adopt integrity management programs similar to existing requirements for gas transmission pipelines on December 4, 2009.

[44] HSPD-7 supersedes PDD-63 (par. 37).

[45] Executive Order 13416, "Strengthening Surface Transportation Security," December 5, 2006.

[46] The PSD was formerly known as the Pipeline Security Program Office. The OSPIE was previously known as the Office of Transportation Sector Network Management and, before that, the Intermodal Security Program Office.

[47] Transportation Security Administration, Pipeline Modal Annex, June 2007, pp. 10-11. G8 = Group of Eight (the United States, the United Kingdom, Canada, France, Germany, Italy, Japan, and Russia).

[48] Department of Transportation (DOT), "Action Taken and Actions Needed to Improve Pipeline Safety," CC-2004- 061, June 16, 2004, p. 21.

[49] Government Accountability Office (GAO), Pipeline Security: TSA Has Taken Actions to Help Strengthen Security, but Could Improve Priority-Setting and Assessment Processes, GAO-10-867, August, 2010, Executive Summary.

[50] Transportation Security Administration, personal communication, November 5, 2010.

[51] Mike Gillenwater, TSA, "Pipeline Security Overview," presented to the Alabama Public Service Commission Gas Pipeline Safety Seminar, Montgomery, AL, December 11, 2007.

[52] Department of Homeland Security, "Extension of Agency Information Collection Activity Under OMB Review: Critical Facility Information of the Top 100 Most Critical Pipelines," 76 Federal Register 62818, October 11, 2011.

[53] Transportation Security Administration, personal communication, February 24, 2012.

[54] TSA, TSA Multi-Modal Criticality Evaluation Tool, TSA Threat Assessment and Risk Management Program, slide presentation, April 15, 2003.

[55] TSA, Intermodal Security Program Office, Pipeline Security Best Practices, October 19, 2005, p. 1.

[56] Transportation Security Administration, Pipeline Security Guidelines, April, 2011, pp. 16-19; Personal communication, February 2, 2010.

[57] U.S. Office of Management and Budget, Budget of the United States Government, Fiscal Year 2012: Appendix, February 2012, p. 508.

[58] Transportation Security Administration, Pipeline Security Division, personal communication, November 5, 2010.

[59] Kip Hawley, Assistant Secretary, Department of Homeland Security, Testimony before the Senate Committee on Commerce, Science, and Transportation hearing on Federal Efforts for Rail and Surface Transportation Security, January 18, 2007.

[60] Pipeline and Hazardous Materials Safety Administration (PHMSA), "Pipeline Safety: Lessons Learned From a Security Breach at a Liquefied Natural Gas Facility," Docket No. PHMSA-04-19856, Federal Register, Vol. 71, No. 249, December 28, 2006, p. 78269; TSA, Intermodal Security Program Office, personal communication, August 30, 2006.

[61] Transportation Security Administration, personal communication, July 6, 2007.

[62] Government Accountability Office, GAO Watchdog, "Transportation Security's Efforts To Ensure Pipeline Security," Assignment No. 440768, Internet database, February 4, 2010.

[63] U.S. Government Accountability Office (GAO), Pipeline Security: TSA Has Taken Actions to Help Strengthen Security, but Could Improve Priority-Setting and Assessment Processes, GAO-10-867, August, 2010, pp. 54-55.

[64] Jerald E. Levine, Director, Departmental GAO/OIG Liaison Office, U.S. Dept. of Homeland Security, Letter to GAO, July 23, 2010.

[65] U.S. Code of Federal Regulations. 18 C.F.R. 157.

[66] Federal Energy Regulatory Commission (FERC), personal communication, May 22, 2003.

[67] Federal Energy Regulatory Commission (FERC), News release, R-01-38, Washington, DC, September 14, 2001.

[68] Federal Energy Regulatory Commission (FERC), Federal Energy Regulatory Commission Annual Report FY2005, 2006, p. 19. These are the most recent specific figures reported.

[69] Federal Energy Regulatory Commission (FERC), Federal Energy Regulatory Commission Annual Report FY2006, 2007, p. 23.

[70] Federal Energy Regulatory Commission (FERC), News release, R-03-08, Washington, DC. February 20, 2003.

[71] Christian Schmollinger, "FERC OKs Emergency Reconstruction," Natural Gas Week, May 13, 2003.

[72] National Transportation Safety Board (NTSB), 2010 Annual Report, NTSB/SPC-11/01, p. iii.

[73] National Transportation Safety Board (NTSB), "Public Hearing: Natural Gas Pipeline Explosion and Fire, San Bruno, CA, September 9, 2010," web page, March 15, 2011, http://www.ntsb.gov/Events/2011/San_Bruno_CA/ default.htm.

[74] National Transportation Safety Board (NTSB), "Pacific Gas and Electric Company Natural Gas Transmission Pipeline Rupture and Fire, San Bruno, CA, September 9, 2010," NTSB/PAR-11/01, August 30, 2011, p.118.

[75] Deborah A.P. Hersman, Chairman, National Transportation Safety Board, "Opening Statement, Pipeline Accident Report – San Bruno, California, September 9, 2010," August 30, 2011.

[76] NTSB, August 30, 2011, p. 123.

[77] Generally, Class 3 locations have 46 or more buildings intended for human occupancy or lie within 100 yards of either a building or outside area of public assembly; Class 4 locations are areas where buildings with four or more stories are prevalent. For precise definitions, see 49 C.F.R. 192.5.

[78] NTSB, August 30, 2011, pp. 128-132.

[79] U.S. Office of Management and Budget, February 2012, p. 1046.

[80] John D. Porcari, Deputy Secretary, U.S. Department of Transportation, Testimony before the House Committee on Transportation and Infrastructure, Hearing on the Enbridge Pipeline Oil Spill in Marshall, MI, September 15, 2010.

[81] Linda Daugherty, Pipeline and Hazardous Materials Safety Administration, Testimony before the National Transportation Safety Board hearing "Natural Gas Pipeline Explosion and Fire, San Bruno, CA, September 9, 2010," March 2, 2011.

[82] Pipeline and Hazardous Materials Safety Administration, personal communication, November 4, 2010.

[83] U.S. Department of Transportation, Pipeline And Hazardous Materials Safety Administration, FY 2011 Budget Request, February 1, 2010, p. 31, http://www.dot.gov/budget /2011/budgetestimates/phmsa.pdf.

[84] National Association of Pipeline Safety Representatives, RE: Request for Waiver of Prior Three Year Average State Expense Component of the Pipeline Safety Grant Program, letter to Mr. Jeffrey D. Weise, Associate Administrator for Pipeline Safety, Pipeline And Hazardous Materials Safety Administration, October 15, 2010, http://www.wutc.wa.gov/ webimage.nsf/web+objects/CCOPS_DOCs_by_Year/$file/NAPSR%20letter%20to%20PH MSA%20dated%2010-15- 09-Waiver%20Request.pdf.

[85] Stephanie Seay, "Budget Woes May Impede Local Pipe Safety Efforts," Gas Daily, November 8, 2010.

[86] The Honorable Cynthia Quarterman, Administrator, Pipeline and Hazardous Materials Safety Administration, Remarks at the Different Pathways to a Common Goal: PIPA, Damage Prevention, & Greater Public Awareness and Involvement Conference, Pipeline Safety Trust, New Orleans, LA, November 5, 2010.

[87] Senator Daniel K. Inouye, opening statement before the Senate Committee on Commerce, Science and Transportation, hearing on the President's FY2006 Budget Request for the Transportation Security Administration (TSA), February 15, 2005.

[88] Congressman Gus M. Billirakis, Remarks before the House Committee on Homeland Security, Subcommittee on Management, Investigations, and Oversight hearing on "Unclogging Pipeline Security: Are the Lines of Responsibility Clear?," Plant City, FL, April 19, 2010.

[89] Department of Homeland Security, "Intent to Request Approval from OMB of One New Public Collection of Information: Pipeline Corporate Security Review," 74 Federal Register 42086, August 20, 2009.

[90] U.S. Government Accountability Office, Transportation Security: Comprehensive Risk Assessments and Stronger Internal Controls Needed to Help Inform TSA Resource Allocation, GAO-09-492, March 2009, p. 30, http://www.gao.gov/new.items/d09492.pdf.

[91] For example, see Hon. William J. Pascrell, Jr., statement at the House Committee on Transportation and Infrastructure, Subcommittee on Highways, Transit and Pipelines, hearing on Pipeline Safety, March 16, 2006.

[92] TSA, Pipeline Security Division, personal communication, November 5, 2010.

[93] Congressman Gus M. Billirakis, April 19, 2010.

[94] The committee print is available at http://homeland.house.gov/sites/homeland.house.gov/files/documents/pdf/ 091411_TS_Subcommittee%20Print.pdf.

[95] John Upton, "Gas Fueled Blaze for Almost Two Hours," San Francisco Examiner, Sept. 15, 2010.

[96] National Transportation Safety Board, Texas Eastern Transmission Corporation Natural Gas Pipeline Explosion and Fire, Edison, New Jersey, March 23, 1994, NTSB/PAR-95/01, January 18, 1995.

[97] U.S. Department of Transportation, Remotely Controlled Valves on Interstate Natural Gas Pipelines, September 1999, p.22.

[98] U.S. Department of Transportation, September 1999, pp. 23-24.

[99] Rich Connell, John Hoeffel and Marc Lifsher, "Lawmakers Move to Impose New Requirements for Pipeline Shutoff Valves," Los Angeles Times, September 14, 2010.

[100] Christina Sames, Vice President, American Gas Association, Remarks at the Different Pathways to a Common Goal: PIPA, Damage Prevention, & Greater Public Awareness and Involvement Conference, Pipeline Safety Trust, New Orleans, LA, November 4, 2010

[101] National Transportation Safety Board (NTSB), Pipeline Special Investigation Report: Evaluation of Accident Data and Federal Oversight of Petroleum Product Pipelines, NTSB/SIR-96/02, 1996, p. 37.

[102] Ibid.

[103] Ibid. p. 39.

[104] 49 CFR 195.452(i)(4)

[105] Pacific Gas and Electric Company, "Re: Updates on Natural Gas Transmission System," Letter to the California Public Utilities Commission, October 25, 2010, p. 2-2, http://www.pge.com/includes/docs/pdfs/about/newsroom/ puc_updates_oct252010.pdf.

[106] Cecil R. Sparks, Thomas R. Morrow, and John P. Harrell, "Cost Benefit Study of Remote Controlled Main Line Valves, Final Report to Gas Research Institute," Southwest Research Institute, San Antonio, TX, Report No. GRI- GRI-98/0076, May 1998.

[107] U.S. Department of Transportation, September 1999, p. 11.

[108] Pacific Gas and Electric Company, October 25, 2010, p. 2-3.

[109] John Upton, "PG&E Faces High Costs on Pipelines," New York Times, March 3, 2011.

[110] Charles H. Batten, "Engineering Analysis of Olympic Pipe Line Company's Safety and Risk Mitigation Features for Application No. 96-1 Cross Cascade Pipeline Project," Batten & Associates, Inc., Locust Grove, VA, February 8, 1999, p. 19, http://www.efsec.wa.gov/oplarchive/oplpft/Cca/chb-3.pdf.

[111] National Transportation Safety Board, Pipeline Accident Summary Report: Natural Gas Pipeline Rupture and Fire During Dredging of Tiger Pass, Louisiana, October 23, 1996, PR98-916501, NTSB/PAR-98/01/SUM, September 28, 1998, p. 15.

[112] See, for example: Tyler Williams, "Cyber Security Threats to Pipelines and Refineries," Pipeline & Gas Journal, November 1, 2007.

[113] U.S. Department of Transportation, September 1999, pp. 19-20.

[114] William M. Nugent, First Vice President, National Association of Regulatory Utility Commissioners, Testimony before the Senate Energy and Natural Resources Committee hearing on Federal, State, and Local Impediments to Siting Energy Infrastructure (May 15, 2001).

[115] National Commission on Energy Policy, Siting Critical Energy Infrastructure: An Overview of Needs and Challenges. (Washington, DC: June 2006): 1. (Hereafter referred to as NCEP 2006.)

[116] Energy Information Administration, Additions to Capacity on the U.S. Natural Gas Pipeline Network: 2005 (August 2006): 11.

[117] Federal Energy Regulatory Commission (FERC), "Commission Approves Revised $1 Billion Millennium Pipeline Project to Bring New Gas Service to the Northeast," Press release (December 21, 2006). See, for example: Randal C. Archibold, "Fighting Plans for a Gas Pipeline: Not Under My Backyard," New York Times (August 7, 2001).

[118] See, for example, Samantha Santa Maria, "Energy Projects: Rockies Express Add-on Pipe Projects Face Several Obstacles to Building in US Northeast," Inside F.E.R.C. (October 22, 2007).

[119] U.S. Department of Transportation, "DOT Issues Much-Anticipated Rules to Enhance Pipeline Safety," Office of Public Affairs, press release, December 3, 2009.

[120] See, for example: Representative James L. Oberstar, Statement before the House Committee on Transportation and Infrastructure, Hearing on the Enbridge Pipeline Oil Spill in Marshall, MI, September 15, 2010.

[121] S.L. Gerard, Pipeline and Hazardous Materials Admin.(PHMSA), Testimony before the House Energy and Commerce Committee, Energy and Air Quality Subcommittee hearing on Pipeline Safety, Serial No. 109-84, April 27, 2006, p. 14.

[122] Ibid. These figures only reflect administrative enforcement cases. They exclude cases that PHMSA has referred to the Department of Justice for civil and criminal enforcement under 49 CFR 190.231 and 190.235.

[123] Pipelines and Hazardous Materials Safety Administration, (PHMSA), "Civil Penalty Cases: Nationwide," Web page, February 8, 2012, http://primis.phmsa.dot.gov/comm/reports/enforce/CivilPenalty_opid_0.html?nocache=9398; "Colorado Pipeline Company Fined 2.3 Million After Explosion," Clean Skies News, December 1, 2009.

[124] PHMSA, February 8, 2012.

[125] Pipeline and Hazardous Materials Safety Administration, Consent Agreement: In the Matter of Kinder Morgan Energy Partners, L.P., Respondent, CPF No. 5-2005-5025H, April 4, 2006.

[126] Kinder Morgan Energy Partners, L.P., "Kinder Morgan Energy Partners Enters into Consent Agreement with PHMSA," press release, Houston, TX, April 10, 2006.

[127] U.S. Environmental Protection Agency, "Kinder Morgan, SFPP Agree To Pay Nearly $5.3 Million To Resolve Federal And State Environmental Violations," press release, May 21, 2007.

[128] U.S. Environmental Protection Agency, "Plains Pipeline to Spend $41 Million to Prevent Oil Spills Across 10,000 Miles of Pipeline," press release, August 10, 2010.

[129] Enbridge Energy Partners, L.P., Enbridge Energy Partners, L.P. Third Quarter 2010 Earnings, Slide presentation, October 28, 2010, p. 8, http://phx.corporate-ir.net/External.File?item=UGFyZW50SUQ9MjY2NzE3N3xDaGlsZElEPTQwMTI5MXxUeXBlPTI=&t=1.

[130] Enbridge Energy Partners, L.P., October 28, 2010, p. 8.

[131] "Olympic Pipe Line, Others Pay Out Record $75 Million in Pipeline Explosion Wrongful Death Settlement," Business Wire, April 10, 2002.

[132] National Transportation Safety Board, Pipeline Accident Report, PAR-03-01, February 11, 2003.

[133] El Paso Corp., Quarterly Report Pursuant to Section 13 or 15(d) of the Securities Exchange Act of 1934, Form 10- Q, for the period ending June 30, 2002, Houston, TX, 2002.

[134] El Paso Corp., 2002.

[135] For further discussion and analysis of Canadian oil sands pipeline issues, see CRS Report R41875, The U.S.-Canada Energy Relationship: Joined at the Well, by Paul W. Parfomak and Michael Ratner.

[136] Canadian Association of Petroleum Producers (CAPP), "Crude Oil: Forecast, Markets & Pipelines," Calgary, AB, June 2010, p. 7, http://www.capp.ca/GetDoc.aspx?DocId=173003; CRS analysis.

[137] See CRS Report R41668, Keystone XL Pipeline Project: Key Issues, by Paul W. Parfomak et al.; U.S. Department of State, "Briefing on the Keystone XL Pipeline," briefing transcript, January 18, 2012, http://www.state.gov/r/pa/prs/ ps/2012/01/181492.htm; TransCanada Corp., "TransCanada Will Re-Apply for a Keystone XL Permit," press release, January 18, 2012.

[138] TransCanada has proposed a pipeline spur from the Keystone XL pipeline to the Bakken oil shale field in Montana, North Dakota and South Dakota, known as the Bakken Marketlink Project.

[139] Anthony Swift, Susan Casey-Lefkowitz, and Elizabeth Shope, "Tar Sands Pipelines Safety Risks," Natural Resources Defense Council, February 2011, p. 6.

[140] Canadian Energy Resources Conservation Board, "ERCB Addresses Statements in Natural Resources Defense Council Pipeline Safety Report," Press release, Calgary, Alberta, February 16, 2011.

[141] GAO, Pipeline Security and Safety: Improved Workforce Planning and Communication Needed, GAO-02-785, August 2002, p. 22.

[142] U.S. Dept. of Transportation, Office of Inspector General, May 21, 2008, p. 6.

[143] Department of Homeland Security (DHS), Transportation Security Administration Fiscal Year 2005 Congressional Budget Justification, Washington, DC, February 2, 2004, p. 20; TSA, Pipeline Security Division, personal communication, February 17, 2009.

[144] American Gas Association (AGA), American Petroleum Institute (API), Association of Oil Pipelines (AOPL), and American Public Gas Association (APGA), joint letter to members of the Senate Commerce Committee providing views on S. 1052, August 22, 2005.

[145] Barrett, T.J. January 18, 2007.

[146] Sammon, John, Transportation Security Administration, Testimony before the House Transportation and Infrastructure Committee, Railroad, Pipelines, and Hazardous Materials Subcommittee hearing on Implementation of the Pipeline Inspection, Protection, Enforcement, and Safety Act of 2006, June 24, 2008.

[147] TSA, Pipeline Security Division, personal communication, February 17, 2009.

[148] National Transportation Safety Board, NTSB Issues Urgent Safety Recommendations As A Result Of Preliminary Findings In San Bruno Pipeline Rupture Investigation; Hearing Scheduled For March," SB-11-01, press release, January 3, 2011.

[149] Ibid.

[150] Linda Daugherty, March 2, 2011.

[151] Dina Cappielo, "What Lies Beneath," Houston Chronicle, November 12, 2006, p. A1.

[152] "Pig" is the common acronym for "pipeline inspection gauge."

[153] Pete Carey, "Pipeline Inspection Not an Exact Science," San Jose Mercury News, October 11, 2010.

[154] The Pipeline Safety Improvement Act of 2002 (P.L. 107-355) directed the DOT to issue regulations on using internal inspection, pressure testing, and direct assessment to natural gas pipelines in high consequence areas.

[155] Christina Sames, Vice President, American Gas Association, November 4, 2010.

[156] John Kiefner, "Overview of Hydrostatic Testing," Presentation to the Hydrostatic Testing Symposium, California Public Utilities Commission, May 6, 2011, http://www.cpuc .ca.gov/NR/rdonlyres/1A47C67C-4398-49CA-B52A-A8B5CD13457B/0/HydrostaticTestingSymposiumPresentationMaterialsversiontopost.pdf.

[157] For an example of such a review, see The Northern Great Plains at Risk: Oil Spill Planning in Keystone Pipeline System, Plains Justice, Billings, MT, November 23, 2010.

[158] Sharon Theimer, "Government Lacks Copies of Emergency Response Plans Developed by Gas Pipeline Operators," Associated Press, October 6, 2010.

[159] Andrew Black, President, Association of Oil Pipelines, Remarks at the Different Pathways to a Common Goal: PIPA, Damage Prevention, & Greater Public Awareness and Involvement Conference, Pipeline Safety Trust, New Orleans, LA, November 5, 2010.

[160] Rob Rogers, "Photo Shows Broken Silvertip Pipeline in Yellowstone River," Billings Gazette, September 7, 2011.

[161] Cynthia L. Quarterman, Administrator, Pipeline and Hazardous Materials Safety Administration, Testimony before the House Committee on Transportation and Infrastructure, Subcommittee on Railroads, Pipelines, and Hazardous Materials Hearing on the Silvertip Pipeline Oil Spill in Yellowstone County, Montana," July 14, 2001, p. 4.

In: America's Pipelines
Editor: Jaron Wiacek

ISBN: 978-1-62417-994-5

Chapter 2

PIPELINE CYBERSECURITY: FEDERAL POLICY*

Paul W. Parfomak

SUMMARY

The vast U.S. network of natural gas and hazardous liquid pipelines is integral to U.S. energy supply and has vital links to other critical infrastructure. While an efficient and fundamentally safe means of transport, this network is vulnerable to cyber attacks. In particular, cyber infiltration of supervisory control and data acquisition (SCADA) systems could allow successful "hackers" to disrupt pipeline service and cause spills, explosions, or fires—all from remote locations. In March 2012, the Department of Homeland Security (DHS) reported ongoing cyber intrusions among U.S. natural gas pipeline operators. These intrusions have heightened congressional concern about cybersecurity in the U.S. pipelines sector.

The Transportation Security Administration (TSA) is authorized by federal statute to promulgate pipeline physical security and cybersecurity regulations, if necessary, but the agency has not issued such regulations. TSA officials assert that security regulations could be counterproductive because they could establish a general standard below the level of security already in place for many pipelines. An April 2011 White House

* This is an edited, reformatted and augmented version of a Congressional Research Service publication, CRS Report for Congress R42660, from www.crs.gov, prepared for Members and Committees of Congress, dated August 16, 2012.

proposal and the Cybersecurity Act of 2012 (S. 2105) both would mandate cybersecurity regulations for privately owned critical infrastructures sectors like pipelines. A revised version of S. 2105, S. 3414, would permit the issuance of regulations but would focus on voluntary cybersecurity measures.

While the pipelines sector has many cybersecurity issues in common with other critical infrastructure sectors, it is somewhat distinct in several ways:

- Pipelines in the United States have been the target of several confirmed terrorist plots and attempted physical attacks since September 11, 2001.
- Changes to pipeline computer networks over the past 20 years, more sophisticated hackers, and the emergence of specialized malicious software have made pipeline SCADA operations increasingly vulnerable to cyber attacks.
- There recently has been a coordinated series of cyber intrusions specifically targeting U.S. pipeline computer systems.
- TSA already has statutory authority to issue cybersecurity regulations for pipelines if the agency chooses to do so, but it may not have the resources to develop, implement, and enforce such regulations if they are mandated.

TSA maintains that voluntary standards have been effective in protecting U.S. pipelines from cyber attacks. Based on the agency's corporate security reviews, TSA believes cybersecurity among major U.S. pipeline systems is effective. However, without formal cybersecurity plans and reporting requirements, it is difficult for Congress to know for certain. Whether the self-interest of pipeline operators is sufficient to generate the level of cybersecurity appropriate for a critical infrastructure sector is open to debate. If Congress concludes that current voluntary measures are insufficient to ensure pipeline cybersecurity, it may decide to provide specific direction to the TSA to develop regulations and provide additional resources to support them, as such an effort may be beyond the TSA Pipeline Security Division's existing capabilities.

INTRODUCTION

Over 500,000 miles of high-volume pipeline gather and transport natural gas, oil, and other hazardous liquids across the United States. In addition, nearly 900,000 miles of smaller distribution pipeline deliver natural gas to businesses and homes.[1] This vast pipeline network is integral to U.S. energy supply and has links to power plants, refineries, airports, and other critical infrastructure. While pipelines are an efficient and fundamentally safe means of transport, many carry volatile, flammable, or toxic materials with the potential to cause public injury and environmental damage. Consequently, pipeline systems have drawn attention as possible targets for terrorism or other malicious activity. Although physical attacks on pipelines have been a focus in North America and elsewhere, the sophisticated computer systems used to operate pipeline systems are also vulnerable to cyber attacks. In particular, cyber infiltration of supervisory control and data acquisition (SCADA) systems could allow "hackers" to disrupt pipeline service and cause spills, explosions, or fires—all from remote locations via the Internet or other communication pathways.

In March 2012, the Department of Homeland Security (DHS) reported ongoing cyber intrusions among U.S. natural gas pipeline operators.[2] The incidents drew new attention to an Al Qaeda video obtained in 2011 by the Federal Bureau of Investigation (FBI) reportedly calling for "electronic jihad" against U.S. critical infrastructure.[3] These cybersecurity events coupled with serious consequences from recent pipeline accidents have heightened congressional concern about cybersecurity measures in the U.S. pipelines sector.

The Transportation Security Administration (TSA) is authorized by federal statute to promulgate pipeline physical security and cybersecurity regulations, if necessary, but the agency has not found a need to issue such regulations to date. An April 2011 White House proposal[4] and the Cybersecurity Act of 2012 (S. 2105) both would mandate the promulgation of cybersecurity regulations for pipelines, among other privately-owned critical infrastructures sectors. A revised version of S. 2105, S. 3414, would permit the issuance of regulations but would focus on voluntary cybersecurity measures. Whether pipelines would be better protected under cybersecurity regulations or under voluntary standards is the subject of ongoing debate.

This report reviews federal programs specifically addressing pipeline cybersecurity, as well as related private sector initiatives and key policy concerns.

Pipeline Security Risks

The pipeline sector has long been considered vulnerable to intentional damage by malicious individuals or organizations—either by physical attacks or by cyber attacks on their computerized SCADA systems.[5] Cybersecurity risks reflect both general threats to the pipeline sector as a whole and specific threats to SCADA systems as the focus of attack. These threats are discussed in the following sections.

General Security Threats to U.S. Pipelines

Since September 11, 2001, federal security warnings have identified pipelines as potential terror targets in the United States.[6] Until recently, attention was most heavily focused on physical threats to pipelines, especially in light of several actual plots involving physical pipeline attacks on U.S. soil. In 2006, for example, federal authorities acknowledged the discovery of a detailed posting on a website linked to Al Qaeda that reportedly encouraged attacks on U.S. pipelines using weapons or hidden explosives.[7] In 2007, the U.S. Department of Justice arrested members of a terrorist group planning to blow up jet fuel pipelines and storage tanks at the John F. Kennedy International Airport in New York.[8] Also in 2007, a U.S. citizen was convicted of trying to conspire with Al Qaeda to attack the Trans Alaska Pipeline System as well as a major natural gas pipeline in the eastern United States.[9] In 2011, federal agents arrested a U.S. citizen for planting an explosive device (which failed to detonate) under a natural gas pipeline in Oklahoma.[10] In June 2012, a man was critically injured attempting to plant an improvised explosive device along a natural gas pipeline in Plano, Texas.[11]

Notwithstanding the security incidents cited above, a January 2011 federal threat assessment concluded "with high confidence that the terrorist threat to the U.S. pipeline industry is low."[12] However, this assertion appears to have taken account of primarily physical threats, because the assessment also stated that "terrorist groups have discussed attacks on unspecified SCADA systems, but it is uncertain whether al-Qa'ida or any other group has the capability to conduct a successful cyber attack on these systems."13 Subsequent events have increased pipeline sector concerns about cyber threats.

SCADA System Security Risks

Supervisory control and data acquisition (SCADA) systems are software-based industrial control systems used to monitor and control many aspects of network operation for railways, utility power grids, water and sewer systems, and pipeline networks. In the pipelines sector, SCADA systems collect data (e.g., line pressure) in real time from sensors throughout a pipeline network, displaying those data to human operators in remote network control rooms. These operators can send computerized commands from SCADA workstations to control geographically dispersed pipeline equipment such as valves, pumps, and compressor stations. The SCADA system provides continuous feedback about conditions all along a pipeline, generating safety alarms when operating conditions fall outside prescribed levels.[14] Communications links throughout the pipeline network may employ dedicated telephone landlines, wireless communications (satellite, microwave, and radio), cellular telephone service, Wi-Fi, and the Internet. As SCADA technology has matured, system control has become more intelligent and more automated, requiring less human intervention.[15]

Historically, pipeline SCADA systems employed highly customized proprietary software and were physically isolated from outside communications and computer networks. Because many of these systems were largely unique to a specific pipeline operator, it would have been difficult for malicious individuals outside the company to access a SCADA system and know what to do with it.

However, due to improvements in computer technology and the ongoing development of communications and Internet-based control system applications, SCADA systems have become much more vulnerable to outside intrusion and manipulation.[16] Specific SCADA security weaknesses include the adoption of standardized control system technologies with known vulnerabilities, increased connection to external networks, insecure communication connections, and the public availability of sensitive information about control systems and infrastructure.[17]

Once accessible to a knowledgeable attacker, a SCADA system can be exploited in a number of specific ways to carry out a pipeline cyber attack:

- issuing unauthorized commands to control equipment;
- sending false information to a control-system operator that initiates inappropriate actions;
- disrupting control system operation by delaying or blocking the flow of information through the control network;
- making unauthorized changes to control system software to modify alarm thresholds or other configuration settings; and
- rendering resources unavailable by propagating malicious software (e.g., a virus, worm, Trojan horse) through the control network.[18]

Depending upon the configuration of a particular pipeline system, such cyber attacks could potentially disrupt pipeline service, damage pipeline equipment (e.g., with excessive pressure), or cause a hazardous release of pipeline commodities into the environment. Even if a hacker did not intend to damage or disrupt the pipeline system, by gaining access to or control of the SCADA system, the intruder could cause serious harm unintentionally.

There have been no major pipeline commodity releases in the United States that investigators have attributed to malicious cyber activity, but SCADA-related problems were a primary cause or contributing factor in several recent pipeline accidents which had catastrophic consequences.

- *San Bruno, CA*—A 2010 natural gas pipeline explosion killed 8 people, injured 60 others, and destroyed 37 homes. Erroneous and unavailable SCADA pressure readings and other SCADA deficiencies were partly responsible for excessive line pressure which ruptured the pipeline.[19]
- *Marshall, MI*—A 2010 pipeline spill released 819,000 gallons of crude oil into a tributary of the Kalamazoo River. Various SCADA control center deficiencies, including the mishandling of pressure alarms, delayed the spill response and increased the size of the spill.[20]
- *Bellingham, WA*—A 1999 gasoline pipeline explosion killed three people and caused $45 million in damage to a city water plant and other property. The SCADA system used to operate the pipeline became unresponsive, making it difficult to analyze pipeline conditions and respond to operational problems that led to the pipeline failure.[21]

While these incidents were all accidental, they are indicative of physical consequences that could result from a pipeline release initiated by a cyber attacker.

Cyber Threats to U.S. Pipelines

In March 2012, the Industrial Control Systems Cyber Emergency Response Team (ICS-CERT) within the Department of Homeland Security (DHS) identified an ongoing series of cyber intrusions among U.S. natural gas pipeline operators dating back to December 2011. According to the agency, various pipeline companies described targeted spear-phishing[22] attempts and intrusions into multiple natural gas pipeline sector organizations "positively identified ... as related to a single campaign."[23] In 2011, computer security company McAfee reported similar "coordinated covert and targeted" cyber attacks originating primarily in China against global energy companies. The attacks began in 2009 and involved spear-phishing, exploitation of Microsoft software vulnerabilities, and the use of remote administration tools to collect sensitive competitive information about oil and gas fields.[24] In 2010, the Stuxnet computer worm was first identified as a threat to industrial control systems. Although the Stuxnet software initially spreads indiscriminately, the software includes a highly specialized industrial process component targeting specific Siemens industrial SCADA systems.[25] The Stuxnet program was reportedly developed thorough a joint United States-Israeli effort to target Iranian computer systems, but it may serve as a model for similar types of malicious programs that could be developed by others in the future.[26] Computer security specialists claim that malicious software developers have already created new software programs tailored to target the kinds of SCADA system weaknesses revealed by Stuxnet.[27]

The increased vulnerability of pipeline SCADA systems due to their modernization, taken together with the emergence of SCADA-specific malicious software and the recent cyber attacks suggests that cybersecurity threats to pipelines have been increasing. Federal agencies and pipeline operators are aware of these threats, however, and have programs in place to counter them. These programs are discussed in the following section.

U.S. PIPELINE CYBERSECURITY INITIATIVES

The federal program for U.S. pipeline security began under the Department of Transportation (DOT)[28] immediately after the terror attacks of September 11, 2001. The Aviation and Transportation Security Act of 2001 (P.L. 107-71), which established the Transportation Security Administration within the DOT, authorized the agency "to issue, rescind, and revise such regulations as are necessary" to carry out its functions (§101). TSA was transferred to the Department of Homeland Security (DHS), newly created under the Homeland Security Act of 2002 (P.L. 107-296). Homeland Security Presidential Directive 7 maintained DHS as the lead agency for pipeline security, and instructed the DOT to "collaborate in regulating the transportation of hazardous materials by all modes (including pipelines)." The Implementing Recommendations of the 9/11 Commission Act of 2007 (P.L.

110-53) directs TSA to promulgate pipeline security regulations and carry out necessary inspection and enforcement if the agency determines that regulations are appropriate (§1557(d)). Thus, TSA has primary responsibility and regulatory authority for the security of natural gas and hazardous liquid (e.g., oil, carbon dioxide) pipelines in the United States. Pipeline security activities at TSA are led by the Pipeline Security Division (PSD) within the agency's Office of Security Policy and Industry Engagement (OSPIE). The PSD is staffed by 13 employees funded out of TSA's general budget.

Although TSA has regulatory authority for pipeline security under P.L. 107-71 and P.L. 110-53, its activities to date have relied upon voluntary industry compliance with the agency's security guidance and best practice recommendations.[29] This guidance includes a number of general cybersecurity provisions.[30] Under its Corporate Security Review program, TSA visits the largest pipeline and natural gas distribution operators to review their security plans and inspect their facilities to evaluate whether each company is following the intent of its security guidance. Pipelines are also included in DHS's multi-modal cybersecurity initiatives, such as its Control System Security Program (CSSP).[31] TSA also has established a public/private partnership-based cybersecurity program supporting the National Infrastructure Protection Plan. The agency's initial focus was on raising cybersecurity awareness and performing outreach to the various transportation modes, including pipelines.[32] The Interstate Natural Gas Association of America maintains its own extensive cyber security guidelines for natural gas pipeline control systems.[33] The American Petroleum Institute likewise maintains an industry standard for oil pipeline control system security.[34] Pipeline operators have scheduled participation in DHS-sponsored control systems cybersecurity training and also participate in the DHS Industrial Control Systems Joint Working Group.[35]

In addition to these efforts, the Department of Energy operates the National SCADA Test Bed Program, a partnership with Idaho National Laboratory, Sandia National Laboratories, and other national laboratories which addresses control system security challenges in the energy sector. Among its key functions, the program performs control systems testing, research and development; control systems requirements development; and industry outreach.[36] Sandia Laboratories also performs authorized, defensive cybersecurity assessments for government, military, and commercial customers through its Information Design Assurance Red Team (IDART) program.[37]

ADEQUACY OF VOLUNTARY PIPELINE CYBERSECURITY

Although Congress has been concerned about various issues related to U.S. pipeline security in recent years, a key subject of debate is the adequacy of TSA's voluntary approach to pipeline security generally and cybersecurity in particular. For example, provisions in the Pipeline Inspection, Protection, Enforcement, and Safety Act of 2006 (P.L. 109-468) required the DOT Inspector General (IG) to "address the adequacy of security standards for gas and oil pipelines" (§23(b)(4)). The 2008 IG's report stated that

> TSA's current security guidance is not mandatory and remains unenforceable unless a regulation is issued to require industry compliance.... [DOT] and TSA will need to conduct covert tests of pipeline systems' vulnerabilities to assess the current guidance as well as the operators' compliance.[38]

Although the IG report did not elaborate on this recommendation, covert testing of vulnerabilities would likely include testing of cybersecurity measures in place to protect pipeline SCADA systems and sensitive operating information such as digital pipeline maps, system design data, and emergency response plans.

Consistent with the IG's recommendation, an April 2011 White House proposal[39] and the Cybersecurity Act of 2012 (S. 2105) both would mandate the promulgation of cybersecurity regulations for pipelines, among other provisions. A revised version of S. 2105, S. 3414, would permit the issuance of regulations but would focus on the use of voluntary cybersecurity measures. S. 3414 would not necessarily confer upon TSA any authority it does not already have to regulate pipeline cybersecurity.

In contrast to the IG's conclusions and the legislative proposals above, the pipeline industry has long expressed concern that security regulations, presumably including cybersecurity, could be "redundant" and "may not be necessary to increase pipeline security."[40] Echoing this sentiment, a DOT official testified in 2007 that enhancing security "does not necessarily mean that we must impose regulatory requirements."[41]

TSA officials have similarly questioned the need for new pipeline security regulations, particularly the IG's call for covert testing of pipeline operator security measures. TSA has argued in the past that the agency is complying with the letter of P.L. 110-53 and that its pipeline operator security reviews are more than paper reviews.[42] TSA officials assert that security regulations could be counterproductive because they could establish a general standard below the level of security already in place at many pipeline companies based on their company-specific security assessments. Because TSA believes the most critical U.S. pipeline systems generally meet or exceed industry security guidance, the agency believes it achieves better security with voluntary guidelines, and maintains a more cooperative and collaborative relationship with its industry partners as well.[43]

The Energy Sector Control Systems Working Group makes related assertions in its *Roadmap to Achieve Energy Delivery Systems Cybersecurity* about the effectiveness of cybersecurity standards alone:

> Although standards may elevate cybersecurity across the energy sector, they do so by requiring the implementation of minimum security measures that set a baseline for cybersecurity across an industry. These minimum security levels may not be sufficient to secure the sector against new and quickly evolving risks. Asset owners compliant with standards may still be vulnerable to cyber intrusion.[44]

Thus, in addition to cybersecurity requirements, pipeline companies may also need appropriate management practices, performance metrics, access to intelligence, and other support measures to maximize the effectiveness of their cybersecurity programs.

Although TSA believes its voluntary approach to pipeline security is adequate, Canadian pipeline regulators have come to a different conclusion. In 2010 the National Energy Board of Canada mandated security regulations for jurisdictional Canadian petroleum and natural gas pipelines, some of which are cross-border pipelines serving export markets in the United States. Many companies operate pipelines in both countries.

In announcing these new regulations, the board stated that it had considered adopting the existing cybersecurity standards "as guidance" rather than an enforceable standard, but "taking into consideration the critical importance of energy infrastructure protection," the board decided to adopt the Standard into the regulations.[45]

Establishing pipeline security regulations in Canada is not completely analogous to doing so in the United States as the Canadian pipeline system is much smaller and operated by far fewer companies than the U.S. system. Nonetheless, Canada's choice to regulate pipeline security may raise questions as to why the United States has not.

TSA Pipeline Cybersecurity Resources

An important consideration in TSA's approach to pipeline security is its staff resources. At its current staffing level of 13 full-time equivalent employees, TSA's Pipeline Security Division (PSD) has limited capabilities. (By comparison, the DOT is authorized for up to 145 federal inspection and enforcement personnel for the U.S. pipeline safety program, and also formally draws upon over 400 state pipeline safety inspectors.) Furthermore, none of the PSD staff have the specialized computer system expertise needed to support more extensive cybersecurity activities, although such expertise is available in other TSA divisions.[46]

Congress has long been concerned about staff resources available to implement the nation's pipeline security program. For example, as one Senator remarked in 2005, "aviation security has received 90% of TSA's funds and virtually all of its attention. There is simply not enough being done to address ... pipeline security."[47] At a congressional field hearing in April 2010, another Member expressed concern that TSA's pipeline division did not have sufficient staff to carry out a federal pipeline security program on a national scale.[48] TSA focuses its security inspections on only the 100 largest pipeline and distribution system operators in an effort to make the best use of its limited resources. However, there are questions as to whether the agency as currently structured could develop and implement new security regulations, conduct rigorous security plan verification, and pose a credible threat of enforcement. Developing specific cybersecurity regulations would pose a particular challenge as the PSD has limited existing capability to do so.

CONCLUSION

While the pipelines sector has many cybersecurity issues in common with other critical infrastructure sectors, it is somewhat distinct in several ways:

- Pipelines in the United States have been the target of several confirmed terrorist plots and attempted physical attacks since September 11, 2001.
- Changes to pipeline computer networks over the past 20 years, more sophisticated hackers, and the emergence of specialized malicious

software have made pipeline SCADA operations increasingly vulnerable to cyber attacks.

- There recently has been a coordinated series of cyber intrusions specifically targeting U.S. pipeline computer systems.
- TSA already has statutory authority to issue cybersecurity regulations for pipelines if the agency chooses to do so, but may not have the resources to develop, implement, and enforce such regulations if they are mandated.

TSA maintains that voluntary cybersecurity standards have been effective in protecting U.S. pipelines from cyber attacks. Based on the agency's corporate security reviews, TSA believes cybersecurity among major U.S. pipeline systems is good. However, without formal cybersecurity plans and reporting requirements, it is difficult for Congress to know for certain. To a great extent, the public must therefore rely on the pipeline industry's self-interest to protect itself from cyber threats. Whether this self-interest is sufficient to generate the level of cybersecurity appropriate for a critical infrastructure sector, and whether imposing formal regulations would be counterproductive, is open to debate. Faced with this uncertainty, legislators are forced to rely upon their own best judgment to reach conclusions about the federal pipelines cybersecurity program. If Congress concludes that current voluntary measures are insufficient to ensure pipeline cybersecurity, it may decide to provide specific direction to TSA to develop regulations and provide additional resources to support them, as such an effort may be beyond the TSA Pipeline Security Division's existing capabilities.

End Notes

[1] Pipeline and Hazardous Materials Safety Administration, "Natural Gas Transmission, Gas Distribution, & Hazardous Liquid Pipeline Annual Mileage," online table, July 3, 2012, http://www.phmsa.dot.gov/ pipeline/library/data-stats. Hazardous liquids primarily include crude oil, gasoline, jet fuel, diesel fuel, home heating oil, propane, and butane. Other hazardous liquids transported by pipeline include anhydrous ammonia, carbon dioxide, kerosene, liquefied ethylene, and some petrochemical feedstocks.

[2] Industrial Control Systems Cyber Emergency Response Team (ICS-CERT), "Gas Pipeline Cyber Intrusion Campaign," ICS-CERT Monthly Monitor, April 2012, p.1, http://www.us-cert.gov/control_systems/pdf/ICSCERT_Monthly_Monitor_Apr2012.pdf.

[3] Jack Cloherty, "Virtual Terrorism: Al Qaeda Video Calls for 'Electronic Jihad'," ABC News, May 22, 2012.

[4] The White House, "Legislative Language, Cybersecurity Regulatory Framework for Covered Critical Infrastructure," April 2011, p.33, http://www.whitehouse.gov/sites/default/files/omb/legislative/letters/law-enforcement-provisionsrelated-to-computer-security-full-bill.pdf.

[5] J.L. Shreeve, "Science & Technology: The Enemy Within," The Independent. London, UK, May 31, 2006, p. 8.

[6] "Already Hard at Work on Security, Pipelines Told of Terrorist Threat," Inside FERC, McGraw-Hill Companies, January 3, 2002.

[7] Wesley Loy, "Web Post Urges Jihadists to Attack Alaska Pipeline," Anchorage Daily News, January 19, 2006.

[8] U.S. Department of Justice, "Four Individuals Charged in Plot to Bomb John F. Kennedy International Airport," Press release, June 2, 2007.

[9] U.S. Attorney's Office, Middle District of Pennsylvania, "Man Convicted of Attempting to Provide Material Support to Al-Qaeda Sentenced to 30 Years' Imprisonment," Press release, November 6, 2007; A. Lubrano and J. Shiffman, "Pa. Man Accused of Terrorist Plot," Philadelphia Inquirer, February 12, 2006, p. A1.

[10] Carol Cratty, "Man Accused in Attempted Bombing of Oklahoma Gas Pipeline," CNN, August 12, 2011.

[11] "Grand Jury Indicts Plano Gas Pipeline Bomb Suspect on Weapons Charge," Associated Press, July 11, 2012.

[12] Transportation Security Administration, Office of Intelligence, Pipeline Threat Assessment, January 18, 2011, p. 3.

[13] Ibid.

[14] National Transportation Safety Board, Supervisory Control and Data Acquisition (SCADA) in Liquid Pipelines, NTSB/SS-05/02, November 29, 2005, pp 1-2.

[15] General Accounting Office (GAO), Critical Infrastructure Protection: Challenges and Efforts to Secure Control Systems, GAO-04-354, March 2004, p. 11.

[16] Tobias Walk, "Cyber-attack Protection for Pipeline SCADA Systems," Pipelines International Digest, January 2012, p. 6; Rose Tsang, Cyberthreats, "Vulnerabilities and Attacks on SCADA Networks," working paper, University of California, Goldman School of Public Policy, 2009, p. 2, http://gspp.berkeley.edu/iths/Tsang_SCADA%20Attacks.pdf.

[17] GAO, 2004, pp. 12-13; Eric Byres, "Next Generation Cyber Attacks Target Oil and Gas SCADA," Pipeline & Gas Journal, February 2012; Robert O'Harrow, Jr., "Cyber Search Engine Exposes Vulnerabilities," Washington Post, June 3, 2102.

[18] Tobias Walk, 2012, pp. 7-8.

[19] National Transportation Safety Board (NTSB), Pacific Gas and Electric Company Natural Gas Transmission Pipeline Rupture and Fire, San Bruno, California, September 9, 2010, NTSB/PAR-11/01, August 30, 2011, pp. 5-12.

[20] National Transportation Safety Board (NTSB). Enbridge, Inc. Hazardous Liquid Pipeline Rupture, Board meeting summary, July 25, 2010, http://www.ntsb.gov/news/events/2012/marshall_mi/index.html.

[21] National Transportation Safety Board (NTSB). Pipeline Rupture and Subsequent Fire in Bellingham, Washington, June 10, 1999, NTSB/PAR-02/02, October 8, 2002, p. 56.

[22] "Spear-phishing" involves sending official-looking e-mails to specific individuals to insert harmful software programs (malware) into protected computer systems; to gain unauthorized access to proprietary business information; or to access confidential data such as passwords, social security numbers, and private account numbers.

[23] Industrial Control Systems Cyber Emergency Response Team (ICS-CERT), "Gas Pipeline Cyber Intrusion Campaign," ICS-CERT Monthly Monitor, April 2012, p.1, http://www.us-cert.gov/control_systems/pdf/ICSCERT_Monthly_Monitor_Apr2012.pdf.

[24] McAfee Foundstone Professional Services and McAfee Labs, Global Energy Cyberattacks:"Night Dragon," white paper, February 10, 2011, p. 3, http://www.mcafee.com/us/resources/white-papers/wp-global-energy-cyberattacksnight-dragon.pdf

[25] Tobias Walk, 2007, p. 7.

[26] David E. Sanger, "Obama Order Sped Up Wave of Cyberattacks Against Iran," New York Times, June 1, 2012.

[27] Eric Byres, February 2012.

[28] The DOT regulates natural gas and hazardous liquid pipelines safety.

[29] Transportation Security Administration (TSA), Pipeline Security Guidelines, April 2011, pp. 16-19, and Pipeline Security Smart Practice Observations, September 19, 2011, pp. 4-8.

[30] For example, TSA's guidance advises operators to "conduct a risk assessment to weigh the benefits of implementing wireless networking against the potential risks for exploitation." TSA, April 2011, p. 18.

[31] Department of Homeland Security, U.S. Computer Emergency Readiness Team, Control System Security Program (CSSP), web page, February 27, 2012, http://www.us-cert.gov/control_systems/index.html.

[32] Jack Fox, General Manager, Pipeline Security Division, Transportation Security Administration (TSA), personal communication, February 22, 2012.

[33] Interstate Natural Gas Association of America (INGAA), Control Systems Cyber Security Guidelines for the Natural Gas Pipeline Industry, Washington, DC, January 31, 2011.

[34] American Petroleum Institute, Pipeline SCADA Security, Second Edition, API Std. 1164,Washington, DC, June 2009.

[35] TSA, February 22, 2012.

[36] U.S. Department of Energy, "National SCADA Test Bed," Web page, July 12, 2012, http://energy.gov/oe/nationalscada-test-bed.

[37] Sandia National Laboratories, "The Information Design Assurance Red Team (IDART)," Web page, 2009, http://idart.sandia.gov/.

[38] U.S. Dept. of Transportation, Office of Inspector General, May 21, 2008, p. 6.

[39] The White House, "Legislative Language, Cybersecurity Regulatory Framework for Covered Critical Infrastructure," April 2011, p. 33, http://www.whitehouse.gov/sites/default/files/omb/legislative/letters/law-enforcement-provisionsrelated-to-computer-security-full-bill.pdf.

[40] American Gas Association (AGA), American Petroleum Institute (API), Association of Oil Pipelines (AOPL), and American Public Gas Association (APGA), joint letter to members of the Senate Commerce Committee providing views on S. 1052, August 22, 2005.

[41] T.J. Barrett, Administrator, Pipeline and Hazardous Materials Safety Administration, Department of Transportation, Testimony before the Senate Committee on Commerce, Science, and Transportation hearing on Federal Efforts for Rail and Surface Transportation Security, January 18, 2007.

[42] John Sammon, Transportation Security Administration, Testimony before the House Transportation and Infrastructure Committee, Railroad, Pipelines, and Hazardous Materials Subcommittee hearing on Implementation of the Pipeline Inspection, Protection, Enforcement, and Safety Act of 2006, June 24, 2008.

[43] Jack Fox, General Manager, Pipeline Security Division, Transportation Security Administration (TSA), Remarks before the Louisiana Gas Association Pipeline Safety Conference, New Orleans, LA, July 25, 2012.

[44] Energy Sector Control Systems Working Group, Roadmap to Achieve Energy Delivery Systems Cybersecurity, September 2011, p. 15.

[45] National Energy Board of Canada, Proposed Regulatory Change (PRC) 2010-01, Adoption of CSA Z246.1-09 Security Management for Petroleum and Natural Gas Industry Systems, File Ad-GA-SEC-SecGen 0901, May 3, 2010, p. 1, https://www.neb-one.gc.ca/ll-eng/livelink.exe/fetch/2000/90463/ 409054/614444/A1S7H7_-_Proposed_Regulatory_Change_(PRC)_2010-01. pdf?nodeid=614556&vernum=0.

[46] TSA, July 25, 2012.

[47] Senator Daniel K. Inouye, opening statement before the Senate Committee on Commerce, Science and Transportation, hearing on the President's FY2006 Budget Request for the Transportation Security Administration (TSA), February 15, 2005.

[48] Congressman Gus M. Billirakis, Remarks before the House Committee on Homeland Security, Subcommittee on Management, Investigations, and Oversight hearing on "Unclogging Pipeline Security: Are the Lines of Responsibility Clear?," Plant City, FL, April 19, 2010.

In: America's Pipelines
Editor: Jaron Wiacek

ISBN: 978-1-62417-994-5

Chapter 3

PIPELINE SAFETY: COLLECTING DATA AND SHARING INFORMATION ON FEDERALLY UNREGULATED GATHERING PIPELINES COULD HELP ENHANCE SAFETY*

United States Government Accountability Office

ABBREVIATIONS

ANPRM	Advanced Notice of Proposed Rulemaking
DOT	Department of Transportation
NAPSR	National Association of Pipeline Safety Representatives
PHMSA	Pipeline and Hazardous Materials Safety Administration
psi	pounds per square inch

* This is an edited, reformatted and augmented version of The United States Government Accountability Office publication, Report to Congressional Requesters, GAO-12-388, dated March 2012.

WHY GAO DID THIS STUDY

Pipelines are a relatively safe mode of transportation for hazardous liquid and natural gas and are regulated by the Department of Transportation's (DOT) Pipeline and Hazardous Materials Safety Administration (PHMSA) and state entities. Included in the nation's pipeline network are an estimated 200,000 or more miles of onshore "gathering" pipelines, which transport products to processing facilities and larger pipelines. (See figure.) Many of these pipelines have not been subject to federal regulation based on their generally rural location and low operating pressures. While incidents involving gathering pipelines regulated by PHMSA have resulted in millions of dollars in property damage in recent years, comparable statistics for federally unregulated gathering pipelines are unknown. This report identifies (1) the safety risks that exist, if any, with onshore hazardous liquid and natural gas gathering pipelines that are not currently under PHMSA regulation and (2) the practices states use to help ensure the safety of these pipelines. GAO surveyed state pipeline safety agencies in all 50 states and the District of Columbia; interviewed officials at PHMSA, state pipeline safety agencies, pipeline companies, and industry associations; and analyzed data and regulations.

WHAT GAO RECOMMENDS

DOT should (1) collect data on federally unregulated hazardous liquid and gas gathering pipelines and (2) establish an online clearinghouse or other resource for sharing information on pipeline safety practices. DOT provided technical corrections on a draft of this report.

WHAT GAO FOUND

While the safety risks of onshore gathering pipelines that are not regulated by PHMSA are generally considered to be lower than for other types of pipelines, PHMSA does not collect comprehensive data to identify the safety risks of unregulated gathering pipelines. In response to a GAO survey, state pipeline safety agencies cited construction quality, maintenance practices, unknown or uncertain locations, and limited or no information on pipeline integrity as among the highest risks for federally unregulated pipelines.

Without data on these risk factors, pipeline safety officials are unable to assess and manage safety risks associated with these pipelines. Furthermore, changes in pipeline operational environments cited in response to GAO's survey and by industry officials could also increase safety risks for federally unregulated gathering pipelines. Specifically, land-use changes are resulting in development encroaching on existing pipelines and the increased extraction of oil and natural gas from shale deposits is resulting in the development of new gathering pipelines, some of which are larger in diameter and operate at higher pressure than older pipelines. PHMSA is considering collecting data on federally unregulated gathering pipelines, but the agency's plans are preliminary, and the extent to which PHMSA will collect data sufficient to evaluate the potential safety risks associated with these pipelines is uncertain.

A small number of state pipeline safety agencies GAO surveyed reported using at least one of five practices that were most frequently cited to help ensure the safety of federally unregulated pipelines. These practices include (1) damage prevention programs, (2) considering areas of highest risk to target resources, (3) safety inspections, (4) public outreach and communication, and (5) increased regulatory attention on operators with prior spills or leaks. However, the sharing of information among states on the safety practices used appears to be limited. Some state and PHMSA officials GAO interviewed had limited awareness of safety practices used by other states. Increased communication and information sharing about pipeline safety practices could boost the use of such practices for unregulated pipelines. However, information targeted at gathering pipelines on PHMSA's website, including relevant safety practices and state activities, is limited.

March 22, 2012

The Honorable John D. Rockefeller IV
Chairman
Committee on Commerce, Science, and Transportation
United States Senate

The Honorable Frank R. Lautenberg
Chairman
Subcommittee on Surface Transportation and Merchant Marine
Infrastructure, Safety, and Security
Committee on Commerce, Science, and Transportation
United States Senate

Pipelines are a relatively safe mode of transportation for hazardous liquid and natural gas. The nation's network of more than 2.5 million miles of pipeline is largely regulated by the Department of Transportation's (DOT) Pipeline and Hazardous Materials Safety Administration (PHMSA) and state entities. Part of this network consists of more than 200,000 miles (estimated) of onshore "gathering" pipelines, which transport hazardous liquid and natural gas products from wells to processing facilities and to larger transmission pipelines. Many of these gathering pipelines, however, have not been subject to PHMSA regulation because they are generally located away from population centers and operate at low pressures.[1] In recent years, incidents involving gathering pipelines regulated by PHMSA have resulted in millions of dollars in property damage. Comparable statistics for federally unregulated gathering pipelines are unknown because PHMSA does not collect such data.

You requested that we review pipeline safety issues related to onshore gathering pipelines not regulated by PHMSA. To do so, we examined (1) the safety risks that exist, if any, with onshore hazardous liquid and natural gas gathering pipelines that are not currently under PHMSA regulation and (2) the practices states use to help ensure the safety of these federally unregulated onshore gathering pipelines.

To identify safety risks that may be associated with gathering pipelines, we reviewed PHMSA safety regulations and analyzed data on pipelines regulated by PHMSA to understand the types of pipeline data currently collected, as well as to compare and analyze accident, injury, fatality, and other trends. We determined that these data were complete, reasonable, and sufficiently reliable for the purposes of this report. For both objectives, we developed and administered a survey to 52 pipeline safety agencies[2] in all 50 states and the District of Columbia to collect information on, among other things, onshore gathering pipelines, the perceived pipeline safety risks associated with those pipelines, and the safety practices used for those pipelines. We received a 100 percent response rate for this survey. We also interviewed officials at PHMSA, selected state pipeline safety agencies, pipeline companies, and industry associations. We conducted site visits to Dallas-Fort Worth, Denver, and Pittsburgh; we selected these locations based on geography, existing pipeline infrastructure, and other factors.

We conducted this performance audit from February 2011 to March 2012 in accordance with generally accepted government auditing standards. Those standards require that we plan and perform the audit to obtain sufficient, appropriate evidence to provide a reasonable basis for our findings and conclusions based on our audit objectives. We believe that the evidence obtained provides a reasonable basis for our findings and conclusions based on our audit objectives.

BACKGROUND

Pipelines transport roughly two-thirds of domestic energy supplies through approximately 2.5 million miles of pipelines throughout the United States. These pipelines carry hazardous liquids and natural gas from producing wells to end users (residences and businesses). Within this nationwide system, there are three main types of pipelines.

- *Gathering pipelines*. Gas gathering pipelines collect natural gas from production areas, while hazardous liquid gathering pipelines collect oil and other petroleum products. These pipelines then typically transport the products to processing facilities, which in turn refine and send the products to transmission pipelines. According to PHMSA officials, traditionally, gathering pipelines range in diameter from about 2 to 12 inches and operate at pressures that range from about 5 to 800 pounds per square inch (psi). These pipelines tend to be located in rural areas but can also be located in urban areas. PHMSA estimates there are 200,000 miles of gas gathering pipelines and 30,000 to 40,000 miles of hazardous liquid gathering pipelines.
- *Transmission pipelines*. Transmission pipelines carry hazardous liquid or natural gas, sometimes over hundreds of miles, to communities and large-volume users (e.g., factories).[3] For natural gas transmission pipelines, compression stations located periodically along the pipeline maintain product pressure. Similarly, pumping stations along hazardous liquid transmission pipelines maintain product flow. Transmission pipelines tend to have the largest diameters and pressures of any type of pipeline, generally ranging from 12 inches to 42 inches in diameter and operating at pressures ranging from 400 to 1440 psi.[4] PHMSA has estimated there are more than 400,000 miles of gas and hazardous liquid transmission pipelines.
- *Distribution pipelines*. Gas distribution pipelines continue to transport natural gas to residential, commercial, and industrial customers, splitting off from transmission pipelines. These pipelines tend to be smaller, sometimes less than 1 inch in diameter, and operate at lower

pressures—0.25 to 100 psi. [5] PHMSA has estimated there are roughly 2 million miles of distribution pipelines, most of which are intrastate pipelines. There are no hazardous liquid distribution pipelines.

In recent years, shale oil and gas exploration and related development has led to the construction of new infrastructure, including gathering pipelines. Shale oil and gas refers to product that is trapped within underground shale formations; these fine-grain sedimentary rocks can be rich sources of oil and natural gas. Over the past decade, improvements in drilling technologies have allowed access to large volumes of shale oil and gas deposits in several states that were previously uneconomical to access.

In terms of fatalities and injuries, pipelines are the safest mode for transporting hazardous liquids and natural gas. From 2004 to 2010, there was an average of about 16 fatalities per year for all incidents reported to PHMSA.[6] In comparison, in 2009, over 3,000 fatalities resulted from incidents involving large trucks carrying freight and about 700 additional fatalities resulted from freight railroad incidents. Yet, pipelines face a number of risks—such as corrosion and excavation damage—that can damage the pipeline's integrity and result in leaks and ruptures. A leak generally occurs with a slow release of product over a small area. A rupture involves the sometimes sudden development of a breach in the pipeline, which may cause hazardous liquids to spill or gas to spark and ignite, resulting in an explosion.[7] Such an incident occurred in San Bruno, California, on September 9, 2010, damaging or destroying over 100 homes and killing 8 people.

PHMSA administers the national regulatory program to ensure the safe transportation of hazardous liquid and gas by pipeline. PHMSA carries out its mission through regulation, [8] national consensus standards, research, education, inspections, and enforcement when safety problems or regulatory violations are found.

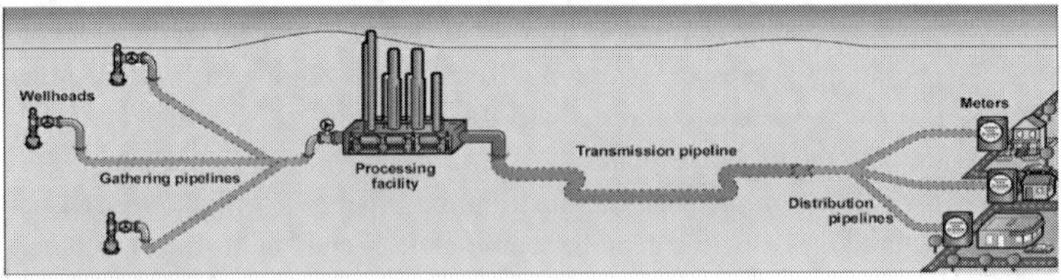

Source: PHMSA.

Figure 1. Pipeline System.

The agency employs over 200 staff in its pipeline safety program, about half of whom inspect hazardous liquid and gas pipelines for compliance with safety regulations. Besides PHMSA, over 300 state inspectors also help oversee pipelines and ensure safety.

In general, PHMSA performs its oversight role using uniform, minimum safety standards that all pipeline operators regulated by PHMSA must meet, as well as a supplemental risk-based regulatory program termed “integrity management” for pipelines in “high-consequence areas” where an incident would have greater consequences for public safety or the environment. The uniform, minimum safety standards include specifications for the design, construction, testing, inspection, operation, and maintenance of pipelines. For example, operators are required to install and maintain pipeline markers to clearly show a pipeline’s right-of-way. In addition, the risk-based integrity management programs for hazardous liquid and natural gas transmission pipelines and natural gas distribution pipelines require operators to systematically identify and mitigate risks to pipeline segments located in high-consequence areas, which are defined differently for three types of pipelines.

- High-consequence areas for hazardous liquid pipelines include areas of highly populated areas, other populated areas, navigable waterways, and areas unusually sensitive to environmental damage.
- High-consequence areas for natural gas transmission pipelines include highly populated or frequently used areas, such as parks.
- Most natural gas distribution pipelines would generally be considered to be in high-consequence areas, as defined under the transmission pipelines regulations, since they are typically located in highly populated areas.

PHMSA regulates hazardous liquid and natural gas gathering pipelines—using uniform, minimum standards—based on their proximity to populated and environmentally sensitive areas. For natural gas gathering pipelines,9 PHMSA uses class locations—the same classification system used for natural gas transmission and distribution pipelines. (See table 1.) Under this system, PHMSA generally regulates onshore natural gas gathering pipelines in Class 2, 3, or 4 locations. For hazardous liquid gathering pipelines, PHMSA regulates those pipelines in incorporated and unincorporated cities, towns, and villages; pipeline segments that cross a waterway currently used for commercial navigation; and certain rural gathering pipelines within one-quarter mile of environmentally sensitive areas. This includes high-consequence areas, as defined for the hazardous liquid integrity management program.

High-consequence areas can also be in Class 1, 2, 3, or 4 locations, which can entail different reporting requirements.

For example, gathering pipeline operators in high-consequence areas that are in Class 1 locations are not required to report data on pipeline-related incidents, including fatality, injury, and property damage information.

Under the current regulatory system, PHMSA does not regulate most gathering pipelines in the United States based on their location. For example, out of the more than 200,000 estimated miles of natural gas gathering pipelines, PHMSA regulates roughly 20,000 miles. Similarly, of the 30,000 to 40,000 estimated miles of hazardous liquid gathering pipelines, PHMSA regulates about 4,000 miles.[10]

However, according to PHMSA officials, the agency has the authority to collect data on all onshore hazardous liquid and gas gathering pipelines,[11] even though it generally does not regulate gas gathering pipelines in Class 1 locations or hazardous liquid gathering pipelines not located in high-consequence areas.

Table 1. PHMSA Class Designations for Gas Pipelines

Class designation	Location features
Class 1	An offshore area or any location with 10 or fewer buildings intended for human occupancy within 220 yards of the centerline of the pipeline.
Class 2	Any location with more than 10 but fewer than 46 buildings intended for human occupancy within 220 yards of the centerline of the pipeline.
Class 3	Any location with more than 46 buildings intended for human occupancy within 220 yards of a pipeline, or an area where the pipeline lies within 100 yards of either a building or a small, well-defined outside area (such as a playground) that is occupied by 20 or more persons at least 5 days a week for 10 weeks in any 12-month period.
Class 4	Any location where unit buildings with four or more stories above ground are prevalent.

Source: 49 C.F.R. § 192.5.

Generally, PHMSA retains full responsibility for inspecting and enforcing regulations on interstate pipelines.

However, states may be authorized to conduct inspections for interstate pipelines, as well as inspections and associated enforcement for intrastate pipelines. States can also promulgate regulations for intrastate pipelines, including gathering pipelines. PHMSA has arrangements with 48 states, the District of Columbia, and Puerto Rico to assist with overseeing interstate, intrastate, or both interstate and intrastate pipelines. These arrangements, in which states act as “agents” for PHMSA, can cover hazardous liquid pipelines only, gas pipelines only, or both (see fig. 2). State pipeline safety offices are allowed to issue regulations supplementing or extending federal regulations, but these state regulations must be at least as stringent as the minimum federal regulations.

If a state wants to issue regulations that apply to pipelines that PHMSA does not regulate, such as unregulated gathering pipelines, it must do so under its own (state) authority.

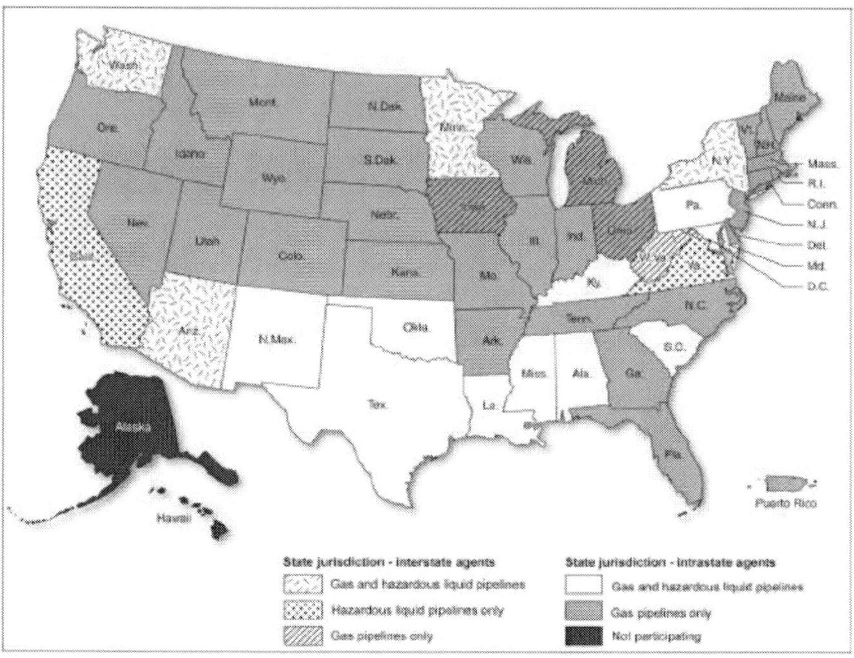

Source: U.S. DOT (information) and Map Resources. (map)

Figure 2. Interstate and Intrastate Agents for PHMSA.

LIMITED INFORMATION ON SAFETY RISKS AND CHANGING OPERATIONAL ENVIRONMENTS ARE LEADING PHMSA TO CONSIDER COLLECTING DATA

Limited Information

While gathering pipelines generally pose lower safety risks than other types of pipelines, PHMSA does not collect comprehensive data on safety risks associated with gathering pipelines. In response to GAO's survey, state pipeline safety agencies cited construction quality, maintenance practices, unknown or uncertain locations, and limited or no information on current pipeline integrity as safety risks for federally unregulated gathering pipelines.[12] Operators of unregulated gathering pipelines are not required by federal law to report information on such risk factors. Consequently, federal and state pipeline safety officials do not know the extent to which individual operators collect such information and use it to monitor the safety of their pipelines. In our survey of 52 state agencies, 39 agencies—10 monitoring hazardous liquid and 29 monitoring natural gas—responded that they had onshore gathering pipelines that PHMSA does not regulate located in their state. (See app. II for a summary of our survey results.) For these 39 agencies, four of the five top responses cited the following risk factors for onshore unregulated gathering pipelines as among the highest public safety risks.

- *Construction quality.* Eighteen state agencies reported that the quality of installation procedures and construction materials is a moderate or high safety risk for unregulated gathering pipelines. The construction phase of pipeline installation is critical to ensure the long-term integrity of the pipeline because the installation methods and materials used in pipeline construction affect the pipeline's resistance to deterioration over time. For example, one inspection requirement for regulated pipelines is that they may not be installed unless they have been visually inspected at the site of installation to ensure that they are not damaged in a manner that could impair their strength or reduce their serviceability.[13] This requirement does not currently apply to unregulated gathering pipelines.
- *Maintenance practices.* Sixteen state agencies reported that the extent to which pipeline operators maintain their pipelines is a moderate or high safety risk for unregulated gathering pipelines. According to

agency officials, after a pipeline is installed and operational, periodic maintenance—such as inspecting and testing equipment—is important to prevent leaks and ruptures and could extend the operating life of a pipeline. Furthermore, preventive measures and repairs conducted on unregulated gathering pipelines, as well as a record of such activities, could provide useful information on the safety and history of a given gathering pipeline.

- *Location.* Sixteen state agencies reported that the unknown or uncertain location of unregulated gathering pipelines presents a moderate or high safety risk. Although individual operators may know the locations of unregulated pipelines, state and local safety agencies may not know or may be uncertain about the locations and mileage of unregulated pipeline infrastructure in their communities. This information is particularly useful for "Call Before You Dig" programs operated by states and localities. If unregulated gathering pipelines are unmarked and program officials do not know the location of the pipelines, businesses and citizens may damage a pipeline during excavation, which could result in an incident—including fatalities, injuries, or damage to property or the environment—as well as the shutting down of the pipeline for repair.
- *Pipeline integrity.* Sixteen state agencies reported that not knowing or having limited knowledge about the integrity—the current condition—of unregulated gathering pipelines is a moderate or high safety risk. Factors that affect the integrity of all pipelines—such as excavation damage and corrosion—also affect gathering pipelines. For example, excavation damage to a pipeline from nearby digging activities (see fig. 3) is the leading cause of pipeline incidents and, as previously noted, the uncertain location of unregulated gathering pipelines may increase the potential for such damage. Furthermore, corrosion can occur on the inside and outside of metal pipelines and is not easily identified without appropriate pipeline assessments. From 2004 through 2010, corrosion was reported as the cause of about 60 percent—or nine incidents—of regulated gas gathering pipeline incidents. Generally, pipeline experts we spoke with said limited information on the integrity of unregulated gathering pipelines prevents analysis to assess the internal and external condition of these pipelines.

Source: Colorado Public Utilities Commission.

Figure 3. Excavation Damage.

Source: U.S. Geological Survey (left) and National Geospatial-Intelligence Agency. (right)

Figure 4. Changing Land Use around Pipelines.

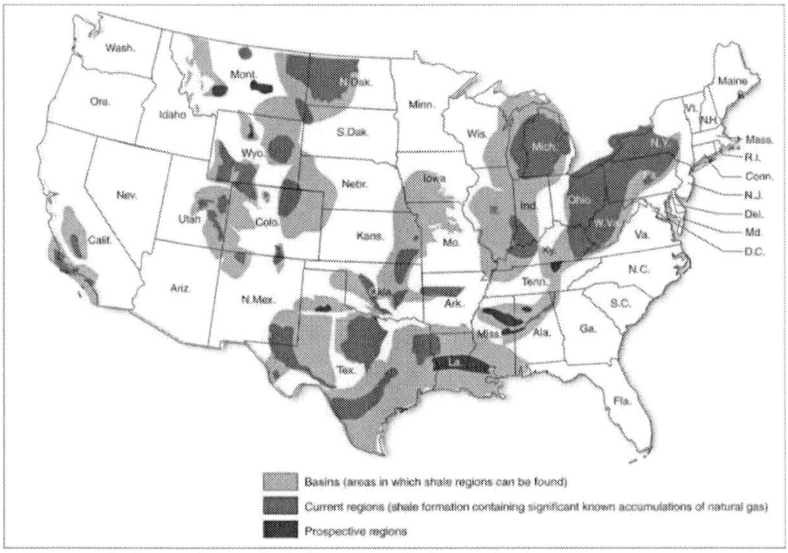

Source: Energy information Administration (information) and Map Recources. (map)
Note: This map depicts shale gas regions, where shale gas is known or thought to exist, but does not indicate where extraction is actually occurring.

Figure 5. Current and Prospective Shale Gas Regions.

Changing Operational Environments

According to responses to our survey and interviews with industry officials and representatives, land-use changes and the increased extraction of oil and natural gas from shale deposits are two changes in the operating environments that could increase the safety risks for unregulated gathering pipelines.

- *Land-use changes.* The fifth top response reported by state pipeline safety agencies we surveyed was that increased urbanization has caused rural areas to become more densely populated and, in some cases, developments have encroached on existing pipeline rights-of-way. (See fig. 4.) Nineteen state agencies reported land-use changes as a moderate or high risk for federally unregulated gathering pipelines. Federal and state pipeline safety officials we spoke with are concerned about the safety and proximity of people who work and live near pipeline rights-of-way. For example, one state official stated that although a new housing or business development can change a location's designation from Class 1 to a higher class that would then fall under PHMSA's jurisdiction, the operator may not be aware of the development and therefore would not monitor and apply more stringent regulations along that pipeline.

- *Increased extraction of oil and gas from shale deposits.* According to pipeline industry officials and representatives we interviewed, the increased extraction of oil and natural gas from shale deposits poses an increased risk to the public, partly because of the development of new and larger gathering pipeline infrastructure. Deposits of oil and natural gas have become increasingly important energy sources in the United States over the past decade (see fig. 5). According to the U.S. Energy Information Administration, shale gas accounted for 16 percent of the total domestic natural gas supply in 2009 and is projected to increase to approximately 47 percent by 2035. This extraction has led to drilling and production in regions of the country that have previously seen little or no such activity. As a result of this ongoing activity, as well as future growth projections, state and federal safety officials we interviewed identified new gathering pipelines related to shale development as a potential public safety risk. The risk is primarily due to the characteristics and quantity of pipeline infrastructure required to support this new production. Specifically, some of these new gathering pipelines have larger diameters and operate at higher pressures that are equivalent to traditional transmission pipelines, but without the regulatory requirements.14 For instance, an October 2010 report15 on pipeline issues and concerns in Fort Worth stated that some gathering pipelines were as large as 24 inches in diameter with maximum allowable operating pressures similar to those for transmission pipelines. Those gathering pipelines were currently exempt from federal integrity management rules, which require some form of pipeline integrity assessment at least once every 7 years, and clearly define how and when problems found during these assessments are to be reported and repaired.

PHMSA Is Considering Collecting Data

PHMSA officials stated that they are considering collecting data on federally unregulated onshore gathering pipelines to better understand and evaluate the safety risks posed by these pipelines. Although PHMSA has the legal authority to collect data on unregulated gathering pipelines, the agency is not required and has not yet exercised its authority to do so. PHMSA officials reported that, instead of collecting such data, the agency was focusing on the development of integrity management requirements and improved data collection for higher-risk transmission and distribution pipelines. However, PHMSA officials reported that there is value in having data for unregulated pipelines similar to what is currently collected on regulated pipelines, such as pipeline characteristics and reportable information on incidents—including the location, cause, and consequences of these incidents.

In addition, PHMSA issued Advanced Notices of Proposed Rulemakings (ANPRM) for onshore hazardous liquid and gas pipelines in October 2010 and August 2011, respectively.[16] For these proposed rulemakings, PHMSA has sought comment on, among other things, whether to extend regulation or other requirements to currently federally unregulated gathering pipelines. Concerning potential data collection, the ANPRMs sought comment on whether to require the submission of annual, incident, and safety-related condition reports on federally unregulated gathering pipelines, as well as on whether to establish a new, risk-based regime of safety requirements for large-diameter, high-pressure gas gathering pipelines, including those pipelines in rural locations.[17] While the ANPRMs did not seek comment on exactly what new data to collect, PHMSA officials reported that the information would likely be similar to what is currently collected on regulated gathering pipelines and that they plan to issue final rules in late 2012. In the event that reporting requirements are adopted, PHMSA officials stated that gathering pipeline data would likely be collected on a state-by-state basis, which could later be expanded to the national level. However, PHMSA's plans for collecting data are preliminary, and the extent to which PHMSA will collect data sufficient to evaluate the potential safety risks associated with these pipelines is uncertain.

Currently, PHMSA collects annual, incident, and safety-related condition data on regulated pipelines. The specific types of safety-related data collected for regulated pipelines include the operator, pipeline system description, mileage by class location, diameter size, operating pressure, incident location, number of injuries and fatalities, property damage, and assessments conducted. These data help federal and state safety officials and pipeline operators increase the safety of these pipelines by better identifying and quantifying safety risks, as well as by implementing mitigation strategies, and addressing potential regulatory needs. It is for these same reasons that PHMSA, state, and some industry officials reported that collecting similar data for unregulated gathering pipelines would be beneficial. PHMSA officials also reported that in the event the agency started collecting data on unregulated onshore gathering pipelines, their current data reporting system could accommodate such a collection and not require large changes for regulators or operators. On the other hand, a few operators and industry groups we met with expressed concerns over the burden that new data reporting would represent. Before any potential data collection reporting requirements could be enacted, PHMSA and the Office of Management and Budget would review and evaluate the value of such information and associated burdens on industry. PHMSA officials said that while many operators should already have information on their gathering pipelines readily available, it would still be important to communicate with operators and take steps to minimize burdens in collecting new gathering pipeline data.

Some benefits of collecting such pipeline data can be seen through additional analysis of currently collected data. For example, PHMSA's data on regulated pipelines indicate that more onshore reportable incidents, as well as total property damage, occur on transmission and distribution pipelines, than on regulated gathering pipelines (see figs. 6 and 7). Although the number of reportable incidents for regulated gas gathering pipelines is lower than for other regulated pipelines, the value of total property damage increased in the past few years. In 2010, these reportable incidents accounted for, on average, about $1.8 million in property damage per incident.

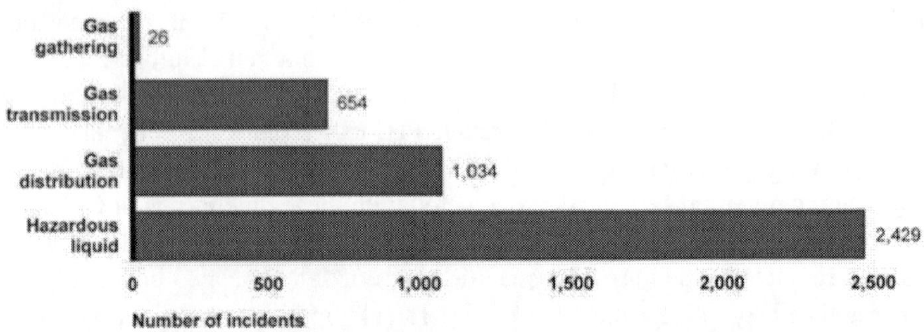

Source: GAO analysis of PHMSA data.

Note: Regulated hazardous liquid data is not collected, segregated, and reported to PHMSA by type of pipeline (i.e., gathering and transmission). Also, there are no hazardous liquid distribution pipeline systems. Therefore, incident data for regulated hazardous liquid transmission and gathering pipelines is combined.

Figure 6. Number of Incidents for All Federally Regulated Pipeline Systems, 2004 through 2010.

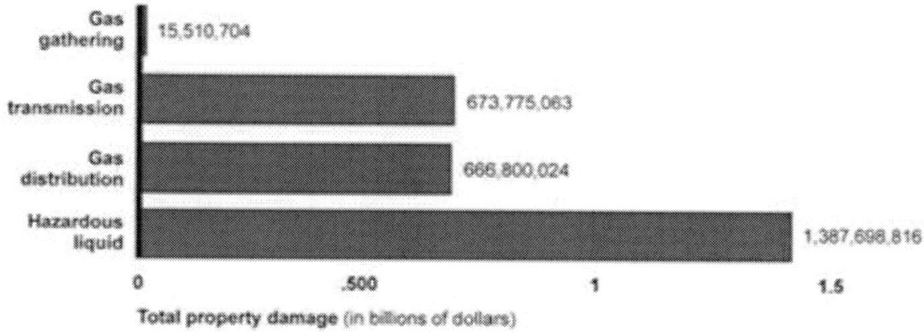

Source: GAO analysis of PHMSA data.

Note: Regulated hazardous liquid data is not collected, segregated, and reported to PHMSA by type of pipeline (i.e., gathering and transmission). Also, there are no hazardous liquid distribution pipeline systems. Therefore, property damage data for regulated hazardous liquid transmission and gathering pipelines is combined.

Figure 7. Total Property Damage for All Regulated Pipeline Systems, 2004 through 2010.

Another benefit of collecting annual, incident, and safety-related condition pipeline data is an increased ability to assess and manage risks. We have previously reported on the importance of assessing and managing risks, including quantifying those risks using data.[18] Data are instrumental in quantifying risks and can reduce uncertainty in assumptions and policy judgments (e.g., safety threats and the likelihood that they will be realized).[19] PHMSA officials reported that collecting data could help to determine the safety risks associated with federally unregulated gathering pipelines, such as tracking injuries, fatalities, and property damage for new gathering pipelines associated with shale development, and whether current safety regulations are appropriate. Related to whether current regulations are appropriate, Congress recently mandated that DOT review the sufficiency of existing federal and state laws and regulations to ensure the safety of hazardous liquid and gas gathering pipelines.[20]

Two industry associations reported that such data collection could help better ensure that federal pipeline programs are appropriately targeted at mitigating safety risks, cost-effective, and not unnecessarily broad in scope. Quantitatively assessing risks could also allow for a ranking and prioritizing of safety risks facing gathering pipelines in a manner that is currently not possible.

Besides PHMSA, states may collect data on unregulated gathering pipelines, but the scope and nature of this data collection can vary. Although the federal government is responsible for setting minimum pipeline safety standards, states can adopt additional or stricter safety standards for intrastate pipeline facilities and transportation—including standards for data collection. For example, Texas's state regulation further defined that the state's safety jurisdiction for onshore gas gathering pipelines begins after the first point of measurement—where the product is first measured to determine the volume being extracted from the well—and is based on population, which is stricter than the federal standard.[21]

Our survey revealed that only 3 of the 39 state agencies reported that they collect and analyze comprehensive pipeline spill and release data on federally unregulated pipelines. Such information can be used to help reduce future incidents.

Additionally, the National Association of Pipeline Safety Representatives (NAPSR) recently conducted a nationwide survey[22] to determine which state requirements match or exceed federal pipeline safety requirements. The survey reported that neither states nor the District of Columbia collected comprehensive data on federally unregulated gathering pipelines, as is required for federally regulated pipelines.

STATES COULD BENEFIT FROM SHARING SAFETY PRACTICES

State pipeline safety agencies reported using five safety practices most frequently to help ensure the safety of onshore hazardous liquid and gas gathering pipelines not regulated by PHMSA, according to our survey of state agencies[23] (see fig. 8). Several of these practices are designed to counter previously discussed safety risks; for example, implementing damage prevention programs can lower the risks of excavation damage.[24] Although these practices were cited most frequently, one-third or less of the state pipeline safety agencies with unregulated gathering pipelines use any one of these practices.

For instance, 13 of the 39 state pipeline safety agencies with unregulated gathering pipelines in their state reported using the most frequently cited safety practice—damage prevention programs. Additionally, some of the state agencies that reported using safety practices also responded that, overall, they had promulgated safety requirements for onshore gathering pipelines that were more stringent than those provided by PHMSA.

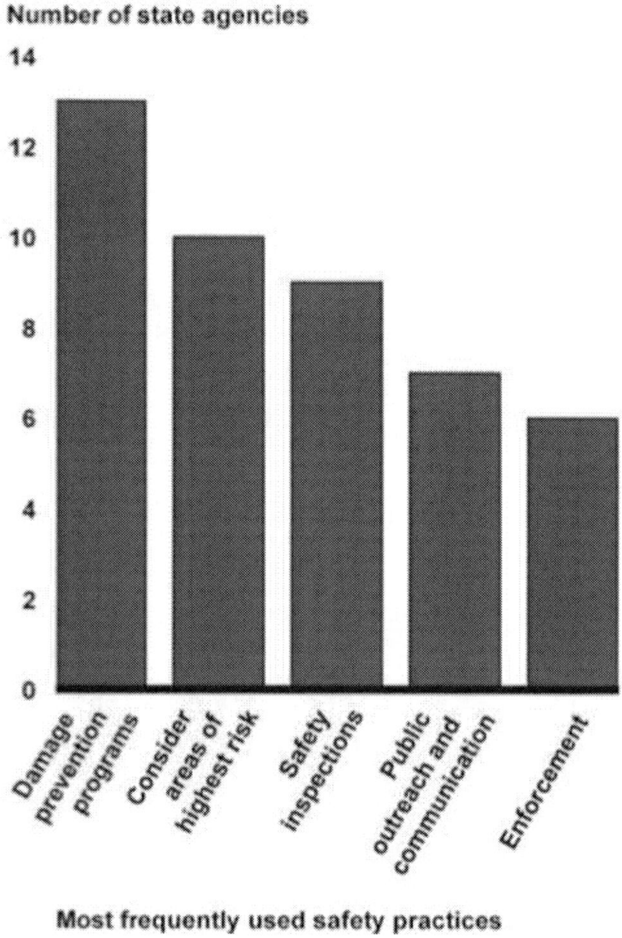

Source: GAO.

Figure 8. Most Frequently Used Safety Practices by States for Gathering Pipelines.

- Damage prevention programs. Thirteen state agencies[25] reported that they implement and enforce a damage prevention program as a practice to help ensure pipeline safety. Damage prevention programs can help mitigate risks and increase safety through a number of activities. For example, damage prevention programs can help reduce the risk of excavation damage by encouraging citizens and other parties to collect information to help identify pipeline locations before digging begins. Damage prevention programs can also include marking the rights-of-way for pipelines—including gathering pipelines—above ground to further reduce the likelihood of excavation damage (see fig. 9). States have developed or participated in damage prevention programs to help reduce instances of excavation damage, including damage to gathering pipelines. For example, Colorado has participated in the national One-Call program to reduce excavation damage. One-Call programs enable citizens and organizations to call an 811 number to notify utilities, pipeline operators, and others about the location and nature of planned digging. Utility, pipeline, or other organization members can then mark where underground pipelines run before any digging begins. Colorado pipeline safety officials reported that some calls related to the marking of regulated and unregulated gathering pipelines. As to the effectiveness of One-Call programs, the Common Ground Alliance has reported that, in 2010, when an excavator notified a call center before digging, damage occurred less than 1 percent of the time.[26]

Source: GAO.

Figure 9. Pipeline Marker.

- *Considering areas of highest risk.* Ten state agencies reported they consider the areas of highest risk to effectively target resources as a safety practice. This approach can help address risks, such as corrosion and a lack of periodic maintenance, by directing oversight to those pipelines that could have the most serious consequences in the event of an incident.[27] In addition, considering the areas of highest risk could help address potential safety risks from new gathering pipeline infrastructure associated with shale development. For example, considering risk factors associated with larger pipelines, operating pressures, and location could help determine the actual risks posed by these new pipelines. Indeed, some of PHMSA's more recent pipeline safety regulations addressing integrity management and high-consequence areas account for risk factors to help determine which regulations might apply to a particular pipeline.28 Industry officials reported that it is more effective to target higher-risk areas than to allocate resources across all areas. Officials with the Texas Oil and Gas Association added that the risk of a pipeline incident in a heavily populated area warrants more attention than the risk of a similar incident in a sparsely populated area. This practice also acknowledges that gathering pipelines run through a wide variety of environments with varying risk levels (see fig. 10). Some states are overseeing pipelines based on identified safety risks. For example, safety compliance and enforcement staff at the Texas Railroad Commission29 reported that inspecting pipeline systems based on identified risks allows the state to inspect some pipelines less frequently, such as pipelines made from newer and safer materials, have advanced monitoring technology, or are located away from populations—like some rural gathering pipelines. Using these risk-based safety evaluations also enables Texas Railroad Commission inspectors to concentrate on higher-risk pipeline systems.
- *Safety inspections.* Nine state agencies reported they conduct recurring, scheduled, or unscheduled safety inspections of hazardous liquid and gas operators as another safety practice. NAPSR officials reported that safety inspections can be regularly scheduled inspections, during which inspectors check system components, specialized inspections (i.e., integrity management) aimed at higher-risk areas, or random checks. These inspections can also help address risks related to the installation and construction quality of a pipeline by ensuring that the pipeline is structurally sound and shows no

evidence of questionable materials or other problems, such as corrosion and excavation damage. PHMSA has recommended that state pipeline safety agencies perform periodic surprise inspections on new pipeline construction to determine whether operators are complying with construction requirements. Inspectors with the Texas Railroad Commission, in addition to sampling on-site pipeline facilities in the field, also review pipeline operators' records and documentation on selected pipeline systems for compliance with federal and state pipeline safety regulations. These risk-based safety evaluations have included the construction of gathering pipelines related to shale development and pipelines not regulated by PHMSA. Such evaluations also help ensure that operators maintain an up-to-date and consistent document records system for installation, operations, and emergency response (see fig. 11).

Source: GAO.

Figure 10. Examples of Rural and Residential Gathering Pipeline Operational Environments near Dallas-Fort Worth.

Source: GAO.

Figure 11. Gathering Pipeline Construction Documentation in Texas.

- *Public outreach and communication.* Seven state agencies reported they engage in outreach or other communication with communities and citizens to boost awareness and knowledge of pipeline safety practices they use. The Common Ground Alliance has reported on the importance of outreach, including the use of structured education programs, targeted mailings, and paid advertising. These and other outreach methods can also underscore the importance of other safety practices, such as damage prevention and One-Call programs. These outreach efforts can involve a number of methods and include educating and engaging the public. In Colorado, Damage Prevention Councils have hosted monthly meetings and participated in local community events—such as educational seminars, parades, and trade shows—to help educate citizens on pipeline safety. Another Colorado entity active in damage prevention is the Colorado Pipeline Association, which comprises pipeline operators dedicated to promoting pipeline safety by providing information for excavators, state residents, businesses, emergency responders, and public officials. In one community, according to a PHMSA official, citizens viewed state safety officials as an objective and neutral party that provided information and perspectives on the planned construction of gathering pipelines. In tandem with private operators, the state officials were able to answer citizen questions and address concerns.

- *Enforcement.* Six state agencies reported a safety practice of establishing a system of escalated enforcement to enhance and increase regulatory attention on operators that have experienced incidents. A pipeline expert we interviewed said that promoting an effective enforcement program was necessary to help ensure pipeline safety. A system of escalated enforcement can enhance and increase regulatory attention on pipeline operators with safety violations. One state pipeline safety official reported that making such attention public can bring additional pressure on and provide incentives for a company to maintain and operate its infrastructure safely. One PHMSA official reported that although many states do not have an enforcement program as elaborate as PHMSA, states with stronger enforcement programs have more of an impact on the operators to increase safety. Pipeline operators may have procedures and established contacts with local enforcement personnel in order to act appropriately to halt dangerous excavation activities that may damage pipelines and potentially cause an immediate threat to life or property. Regarding federally unregulated gathering pipelines, one Colorado official reported that because gathering pipeline companies operate pipelines and conduct excavation work, they would be subject to any necessary enforcement due to safety violations.

However, sharing of information among states on the safety practices they use for unregulated gathering pipelines appears to be limited. Some state and PHMSA officials we interviewed had limited awareness of what other states were doing to help ensure the safety of gathering pipelines not regulated by PHMSA. For example, pipeline safety officials we interviewed had limited awareness of other state programs—sometimes even for an adjacent state—even if those programs were intended to address common risks, such as reducing excavation damage and corrosion. PHMSA officials were likewise unable to report on the safety practices that many states use or on states' regulations that were more stringent than federal requirements. PHMSA's website holds a wealth of information on various pipeline safety topics, including recent pipeline forums and industry research, incident investigations, and other information. However, information targeted at gathering pipelines, including relevant safety practices and state activities, is limited.30 In addition, all related information could be grouped to decrease time spent searching and scanning.

Currently, there is no central PHMSA web page or resource for gathering pipelines, regulated or unregulated—possibly due, in part, to the lower safety risks that regulated gathering pipelines have posed to people and property when compared with other pipelines, like transmission pipelines. PHMSA officials said that its website also focuses on pipelines that PHMSA regulates but excludes most gathering pipelines. PHMSA has considered the development of a website to help facilitate sharing information among states. While this project is still in the planning stages and not targeted at gathering pipelines, it could be a resource to share program and safety practices among states and PHMSA.

Increased communication and information sharing about pipeline safety practices could boost the use of those practices in states with unregulated gathering pipelines. As previously discussed, even the safety practice that our survey respondents reported using most frequently—implementing a damage prevention program—was used by just 13 of the 39 responding state pipeline safety agencies with unregulated gathering pipelines in their state. The other four safety practices cited are reportedly used even less. Improved information sharing among states and PHMSA could help spread information on how these safety practices—which are also used for regulated pipelines—could be applied to unregulated gathering pipelines, thereby benefiting other states with unregulated gathering pipelines. We have previously reported on the value of organizations reporting and sharing safety information as part of encouraging a wider safety culture.31 Safety culture can include organizational awareness of safety and open communication.

The benefits of a strong safety culture have widespread applicability, including in other transportation areas— such as aviation and transit. PHMSA could serve to facilitate feedback and evaluate safety information related to unregulated gathering pipelines in states. By collecting information on safety practices and other information relevant to unregulated gathering pipelines, PHMSA could increase the potential for identifying systemic issues, disseminating lessons learned, and improving pipeline safety across the country. PHMSA officials reported that, in the past, similar online and educational efforts in other areas have resulted in increasing education and information sharing among state pipeline safety officials.

Conclusion

While the safety risks of federally unregulated, onshore hazardous liquid and gas gathering pipelines are generally considered to be lower than other types of pipelines, PHMSA is currently not able to determine the performance and safety of these gathering pipelines because it does not collect the necessary pipeline operator data. The agency is considering options to collect such information, which could facilitate quantitatively assessing the safety risks posed by unregulated gathering pipelines. Furthermore, these data would be critical in helping PHMSA to evaluate the sufficiency of safety regulations for gathering pipelines as required by the congressional mandate or that increasing shale development across the country might necessitate. Making data-driven, evidence-based decisions about the risks of federally unregulated gathering pipelines is especially important in a time of limited resources.

The absence of an information-sharing resource focused on federally unregulated gathering pipelines means that both states and PHMSA could miss opportunities to share lessons learned and successful practices for helping to ensure pipeline safety. Sharing such lessons and related safety reporting can help support a safety culture and increase state officials' awareness of possible safety practices or strategies that they can use to enhance pipeline safety. Lessons learned can also help states avoid the mistakes of others. Additionally, increased information sharing through such a resource would help PHMSA become more aware of state pipeline safety practices and initiatives—which in turn would assist PHMSA in sharing and supporting these safety practices, as well as in considering what state efforts may have applicability for federal programs, regulation, and guidance.

Recommendations for Executive Action

To enhance the safety of unregulated onshore hazardous liquid and gas gathering pipelines, we recommend that the Secretary of Transportation direct the PHMSA Administrator to take the following two actions:

- Collect data from operators of federally unregulated onshore hazardous liquid and gas gathering pipelines, subsequent to an analysis of the benefits and industry burdens associated with such data collection. Data collected should be comparable to what PHMSA collects annually from operators of regulated gathering pipelines (e.g., fatalities, injuries, property damage, location, mileage, size, operating pressure, maintenance history, and the causes of incidents and consequences).

- Establish an online clearinghouse or other resource for states to share information on practices that can help ensure the safety of federally unregulated onshore hazardous liquid and gas gathering pipelines. This resource could include updates on related PHMSA and industry initiatives, guidance, related PHMSA rulemakings, and other information collected or shared by states.

AGENCY COMMENTS

We provided the Department of Transportation with a draft of this report for review and comment. The department provided technical corrections, which we incorporated as appropriate.

Susan A. Fleming
Director
Physical Infrastructure Issues

APPENDIX I: OBJECTIVES, SCOPE, AND METHODOLOGY

The objectives of our review were to determine (1) the safety risks that exist, if any, with onshore hazardous liquid and natural gas gathering pipelines that are not currently under the Pipeline and Hazardous Materials Safety Administration (PHMSA) regulation and (2) the practices states are using to help ensure the safety of unregulated onshore gathering pipelines. To address our objectives, we reviewed PHMSA and other federal agency regulations, as well as available safety data on regulated pipelines. We also interviewed officials at PHMSA, state pipeline safety agencies, pipeline companies and other industry stakeholders, and related associations. We obtained data on pipelines regulated by PHMSA to understand the types of pipeline data currently collected, as well as to compare and analyze accident, injury, fatality, and other trends. We reviewed the data and conducted follow-up work as necessary to determine that the data were complete, reasonable, and sufficiently reliable for the purposes of this report. We also conducted site visits—selecting locations based on geography, pipeline infrastructure, and other factors—to interview pipeline officials and representatives in Denver, Pittsburgh, and Dallas-Fort Worth. We later identified an initial list of safety risks and safety practices through information collection and document review processes.

To determine what safety risks may be associated with federally unregulated gathering pipelines—in addition to reviewing federal agency regulations, regulated pipeline safety data, and conducting various interviews—and because of the lack of historical and nationwide data, we developed and administered a web-based survey to state pipeline safety agencies in all 50 states and the District of Columbia.[1] Our survey was intended to collect information otherwise not available from PHMSA, states, industry, or other sources on safety risks associated with onshore, federally unregulated hazardous liquid and gas gathering pipelines and related safety practices to help address those risks and ensure safety. We used the survey to identify which states had unregulated, onshore gathering pipelines and what perceived pipeline safety risks were associated with those pipelines. To identify safety practices states are using, we reviewed industry documents and conducted interviews with public and private experts and officials. Then, as part of our survey of state pipeline safety agencies, we asked officials to identify the practices they used to ensure the safety of onshore, federally unregulated hazardous liquid and gas gathering pipelines. From our survey results, we identified the most frequently cited safety practices, including additional state programs, activities, and other practices.

To develop the survey questions, we conducted initial interviews with state officials and other pipeline safety stakeholders to identify safety issues regarding unregulated gathering pipelines. We also reviewed key literature to ascertain pipeline safety practices and other issues. We consulted with PHMSA officials and reviewed PHMSA documentation to identify the proper terminology for use in the survey.

The survey was pretested with potential respondents from state pipeline safety agencies, as well as with the Congressional Research Service and National Association of Pipeline Safety Representatives.[2] We did this to ensure that (1) the questions were clear and unambiguous, (2) the terms we used were precise, (3) the survey did not place an undue burden on the agency officials completing it, and (4) the survey was independent and unbiased. In addition, the survey was reviewed by an internal, independent survey expert. We took steps in survey design, data collection, and analysis to minimize nonsampling errors. For example, we worked with PHMSA officials to identify the appropriate survey respondents—state pipeline safety agencies. To minimize measurement error that might occur from respondents interpreting our questions differently from our intended purpose, we extensively pretested the survey and followed up with nonresponding units and with units whose responses violated certain validity checks. We identified only two cases where the respondents had slightly varied responses from our intended question, although the majority understood our questions as intended. Finally, to eliminate data-processing errors, we independently verified the computer program that generated the survey results. Our results are not subject to sampling error because we administered our survey to all 50 state pipeline safety agencies and the District of Columbia.

The survey was conducted using self-administered electronic questionnaires posted on the World Wide Web. We sent e-mail notifications to 52 agencies responding to our survey. We also e-mailed each potential respondent a unique password and username to ensure that only members of the target population could participate in the survey. To encourage respondents to complete the survey, we sent an e-mail reminder to each nonrespondent about 2 weeks after our initial e-mail message. The survey data were collected from July through September 2011. We received responses from all 50 states and the District of Columbia, for an overall response rate of 100 percent. This "collective perspective" obtained from each of the agencies helps to mitigate individual respondent bias by aggregating information across the range of different viewpoints. For purposes of characterizing the results of our survey, we identified specific meanings for the words we used to quantify the results, as follows: "a few" means between 1 percent and 24 percent of respondents, "some" means between 25 percent and 44 percent of respondents, "about half" means between 45 percent and 55 percent of respondents, "a majority" means between 56 percent and 74 percent of respondents, "most" means between 75 percent and 94 percent of respondents, and "nearly all" means 95 percent or more of respondents. This report contains the central results from the survey (see app. II).

We conducted this performance audit from February 2011 to March 2012 in accordance with generally accepted government auditing standards. Those standards require that we plan and perform the audit to obtain sufficient, appropriate evidence to provide a reasonable basis for our findings and conclusions based on our audit objectives. We believe that the evidence obtained provides a reasonable basis for our findings and conclusions based on our audit objectives.

APPENDIX II: SUMMARY RESULTS, GAO PIPELINE SAFETY REGULATIONS SURVEY

General Pipeline Safety Regulation Survey Questions	Hazardous Liquid	Natural Gas
YES RESPONSES	Frequency	Frequency
Does your state have any onshore gathering pipelines outside of high consequence areas that PHMSA does not regulate?	10	29
Does your agency collect any data for onshore gathering pipelines that PHMSA does not regulate?	4	7
Does your state have safety requirements for onshore gathering pipelines that are more stringent than those provided by PHMSA?	1	7
Subpopulation Total	18	52
How great a safety risk, if at all, are the following factors for onshore hazardous liquid and gas gathering pipelines in your state that PHMSA does not regulate?	Hazardous Liquid	Natural Gas
MODERATE AND HIGH SAFETY RISK RESPONSES	Frequency	Frequency
A. Limited or no annual reporting data (similar to PHMSA's) available on these pipelines (e.g., mileage, leaks)	1	9
B. Limited or no incident data available on these pipelines (e.g., spills, releases)	2	11
C. Limited or no information on the integrity of these pipelines[a]	2	13
D. Unknown or uncertain locations of pipelines	3	13
E. Location of these pipelines in high consequence areas[b]	2	7
F. Limited or no inspections conducted on these pipelines[a]	3	13
G. Limited or no information on the pipe size	2	7
H. Limited or no information on operating pressurea		
I. Installation/construction quality	3	15
J. Periodic maintenance not conducted on these pipelines	3	13
K. Quality of product (sour or non-sour, corrosive, abrasive, etc.)	3	11
L. Limited information on reporting damages	2	9
M. Other (Please specify)	1	1
Subpopulation Total	10	29

[a] This response category was combined to represent an overall category of "Limited or no information of the pipeline integrity."

[b] A survey follow up found that the majority of respondents also understood the question as it was intended to mean Class 1 unregulated gathering pipelines previously located in rural areas that may have transitioned to be located in either (Class 2, 3 or 4) high consequence areas due to changing land use.

Does your agency use any of the following practices to ensure onshore hazardous liquid and gas pipeline safetyin your state?	Hazardous Liquid		Natural Gas	
	Regulated Pipelines	Unregulated Gathering	Regulated Pipelines	Unregulated Gathering
YES RESPONSES	Frequency	Frequency	Frequency	Frequency
A. Considering the areas of highest risk to effectively target resources	14	3	47	7
B. Collecting and analyzing comprehensive pipeline spill/release data on various types of pipelines	6	1	27	2
C. Standardizing spill/ release data collection in order to better assess trends or common causes of spills/ releases so that prevention measures can be targeted and evaluated to reduce future incidents	5	1	22	2
D. Engaging in outreach or other communication with communities and citizens to boost awareness or knowledge of pipelines, including their locations	10	3	37	4
E. Coordinating with operators to understand new technologies, including computer software, pipeline inspection devices, and other tools to report on pipelines	11	2	34	2
F. Using existing or emerging technologies to reduce the time required to detect pipeline leaks	8	1	21	3
G. Working to increase construction quality to ensure a long life for pipeline infrastructure	11	2	40	3
H. Prioritizing the replacement or repair of aging or otherwise limited infrastructure	9	1	37	2
I. Conducting recurring, scheduled or unscheduled safety inspections of operators	14	3	47	6
J. Establishing a system of escalated enforcement toenhance and increase regulatory attention on operators that have spills/ releases	11	2	29	4
K. Developing standards or other	7	2	28	2

Does your agency use any of the following practices to ensure onshore hazardous liquid and gas pipeline safetyin your state?	Hazardous Liquid		Natural Gas	
	Regulated Pipelines	Unregulated Gathering	Regulated Pipelines	Unregulated Gathering
YES RESPONSES	Frequency	Frequency	Frequency	Frequency
bench marks to measure and evaluate performance in encouraging safety				
L. Monitoring implementation of corrective or preventive measures to evaluate their impact or effectiveness	13	0	40	4
M. Using existing integrity management practices for PHMSA unregulated lines	8	2	15	2
N. Implementing and enforcing a damage prevention program	13	3	38	10
O. Other (Please specify)	0	1	7	2
Subpopulation Total	18	10	52	29

Source: GAO.

End Notes

[1] PHMSA has limited statutory authority to regulate such pipelines under 49 U.S.C. § 60101(b).

[2] Two state pipeline safety officials from separate agencies in Arkansas that are responsible for overseeing pipeline safety in that state responded to our survey. Also, Alaska and Hawaii have chosen not to participate in pipeline arrangements with PHMSA. Therefore, a PHMSA official who conducts state pipeline safety inspections in these states responded to those states' surveys.

[3] For the purposes of this report, we use the term transmission pipeline to refer to both hazardous liquid and natural gas pipelines carrying product over long distances to users.

[4] However, there are transmission pipelines smaller than 12 inches in diameter and other pipelines that operate at pressures greater than 1440 psi. In addition, we have reported on safety regulations for certain transmission pipelines that operate at lower stress. For more information, see GAO, Safety Effects of Less Prescriptive Requirements for Low-Stress Natural Gas Transmission Pipelines Are Uncertain, GAO-12-389R (Washington, D.C.: Feb. 16, 2012).

[5] However, some distribution pipelines can be as large as 24 inches in diameter and operate at higher pressures (i.e., over 350 psi).

[6] In its regulations, PHMSA refers to the release of natural gas from a pipeline as an "incident" and a spill from a hazardous liquid pipeline as an "accident." (49 CFR Part 195, Subpart B). For simplicity, this report will refer to both as "incidents." PHMSA tracks summary level statistics for pipeline incidents that meet certain reporting criteria, such as those incidents resulting in a fatality.

[7] The risks and consequences posed by gas and hazardous liquids incidents also differ. Gas tends to ignite more easily, resulting in more explosions. Hazardous liquids ignite less easily, but can spill onto and pollute the environment.

[8] Part 191 (Gas Reporting), Part 192 (Gas), Part 193 (Liquid Natural Gas), Part 194 (Liquid Facility Response Plans), and Part 195 (Hazardous Liquid) of Title 49 of the Code of Federal Regulations.

[9] 49 C.F.R. §192.5.

[10] According to PHMSA officials, Alaska, California, Louisiana, and Oklahoma have the majority of federally unregulated gathering pipeline mileage in the United States.

[11] 49 U.S.C. §60102(b).This report refers to gathering pipelines that are not regulated by PHMSA but that PHMSA is not prohibited by law from regulating. Nonregulated gathering pipelines are typically unregulated Class 1 pipelines found in rural locations where there are no or low populations.

[12] Risk is defined by PHMSA as the combination of the likelihood and the consequence of a specified hazard being realized.

[13] 49 C.F.R. §195.206 (Hazardous Liquid) and 49 C.F.R. §192.307 (Gas).

[14] 49 C.F.R. § 195.452 (Hazardous Materials) & 49 C.F.R. Part 192 Subpart O (Gas).

[15] Accufacts Inc. and Pipeline Safety Trust, The State of Natural Gas Pipelines in Fort Worth, a special report prepared at the request of the Fort Worth League of Neighborhoods, Fort Worth, October 2010.

[16] 75 Fed. Reg. 63774 (Oct. 18, 2010) and 76 Fed. Reg. 53086 (Aug. 25, 2011).

[17] PHMSA granted extensions to the comment period for both of these ANPRMs (see 76 Fed. Reg. 303 (Jan. 4, 2011) and 76 Fed. Reg. 70953 (Nov. 16, 2011)). The ANPRM related to hazardous liquids closed for comment in February 2011 and the ANPRM for gas in January 2012.

[18] GAO, Risk Management: Strengthening the Use of Risk Management Principles in Homeland Security, GAO-08-904T (Washington, D.C.: June 25, 2008).

[19] GAO, Homeland Security: DHS Risk-Based Grant Methodology Is Reasonable, But Current Version's Measure of Vulnerability is Limited, GAO-08-852 (Washington, D.C.: June 27, 2008)

[20] Pipeline Safety, Regulatory Certainty, and Job Creation Act of 2011, Pub. L. No. 112-90, §21, 125 Stat. 1904, 1917 (2011).

[21] Texas Administrative Code, Title 16, Part 1, Chapter 8 Subchapter A, Rule Section 8.5. Although the federal government is responsible for setting minimum pipeline safety standards, Texas, and other certified states, can adopt additional or more stringent safety standards for intrastate pipeline facilities and intrastate pipeline transportation. (49 U.S.C. §60104(c)). However, any intrastate safety standards adopted by Texas and other certified states must be compatible with the federal standards.

[22] National Association of Pipeline Safety Representatives and the National Association of Regulatory Utility Commissioners, Compendium of State Pipeline Safety Requirements & Initiatives Providing Increased Public Safety Levels compared to Code of Federal Regulations, 1st Edition 2011 (Sept. 30, 2011).

[23] These practices refer to state pipeline safety practices that are more stringent than PHMSA's regulatory requirements and to central activities, programs, or other policies that can help ensure the safety of various onshore pipelines. We recognize that in some states a practice may be part of the state's regulations, while in other states, such practices are separate from regulations.

[24] When we discuss safety practices, we refer to those safety practices that state agencies described as being relevant for helping ensure the safety of gathering pipelines. Some of these same practices may also be applied to other types of pipelines.

[25] When discussing the number of state agencies that reported using a given safety practice, these 13 agencies are a subset of the 39 state agencies that reported having unregulated gathering pipelines.

[26] Common Ground Alliance, DIRT Annual Report for 2010 (Alexandria, VA, 2011). The Common Ground Alliance is an association dedicated to ensuring public safety, environmental protection, and the integrity of services by promoting effective damage prevention practices.

[27] While data are helpful for assessing risks, PHMSA and state agency officials mentioned using other information sources to qualitatively assess risks, such as guidance and protocols and general pipeline traits like location.

[28] PHMSA's integrity management program for transmission pipelines already requires operators to systematically manage risks to their pipelines in areas where an accident could have the highest consequences. In our past work on pipeline safety, we have supported the use of risk-based methodologies for PHMSA's regulation of the pipeline industry. See GAO, Natural Gas Pipeline Safety: Risk-Based Standards Should Allow Operators to Better Tailor Reassessments to Pipeline Threats, GAO-06-945 (Washington, D.C.: Sept. 8, 2006).

[29] Formed in 1891, the Texas Railroad Commission regulates the safety of intrastate natural gas pipelines and hazardous liquids pipelines in Texas. Commission staff conduct safety inspections of pipeline operators, issue compliance guidance, and enforce rules for Texas's pipeline excavation damage prevention program.

[30] Websites that are limited tend to be less usable. Usable websites—that is, those that are planned and designed to collect data on what users need and include various user-centered design methods, among other things—ensure that needed information is available and clearly displayed. These standards are based on guidance posted on the U.S. Department of Health and Human Services' websites, http://www.usability.gov, a source for information on usability and user-centered design. It provides guidance and tools on how to make websites and other communication systems more usable and useful.

[31] GAO, Rail Transit: FTA Programs Are Helping Address Transit Agencies' Safety Challenges, but Improved Performance Goals and Measures Could Better Focus Efforts, GAO-11-199 (Washington, D.C.: Jan. 31, 2011).

End Notes for Appendix I

[1] Alaska and Hawaii have chosen not to participate in pipeline arrangements with PHMSA. Therefore, a PHMSA official who conducts state pipeline safety inspections in these states responded to those states' surveys.

[2] The National Association of Pipeline Safety Representatives is a national association representing the state pipeline safety inspectors in the contiguous United States, as well as the District of Columbia and Puerto Rico.

In: America's Pipelines
Editor: Jaron Wiacek

ISBN: 978-1-62417-994-5

Chapter 4

SAFETY EFFECTS OF LESS PRESCRIPTIVE REQUIREMENTS FOR LOW-STRESS NATURAL GAS TRANSMISSION PIPELINES ARE UNCERTAIN*

United States Government Accountability Office

The Honorable John D. Rockefeller
Chairman
Committee on Commerce, Science, and Transportation
United States Senate

The Honorable Frank Lautenberg
Chairman
Subcommittee on Surface Transportation
and Merchant Marine Infrastructure, Safety, and Security
Committee on Commerce, Science, and Transportation
United States Senate

More than 2.6 million miles of pipelines form a nationwide network to transport the majority of natural gas consumed in the United States. This

---

* This is an edited, reformatted and augmented version of The United States Government Accountability Office publication, GAO-12 389R, dated February 16, 2012.

extensive gas pipeline network includes several different types of pipelines, including

- *transmission pipelines,* which transport product over long distances at high pressure from sources to communities;
- *distribution pipelines*, which operate at lower pressures to deliver natural gas to homes and businesses; and
- *low-stress transmission pipelines*, which typically connect transmission pipelines to distribution pipelines and may operate at pressures that are similar to the latter.

The Pipeline and Hazardous Materials Safety Administration (PHMSA), within the Department of Transportation (DOT), is responsible for establishing safety requirements for these pipelines and has traditionally included low-stress transmission pipelines in its regulations for all transmission pipelines. In 2004, PHMSA implemented a risk-based regulatory approach called "integrity management" for natural gas transmission pipelines, including low-stress transmission pipelines.[1] In 2010, PHMSA extended integrity management to distribution pipelines, but modified the requirements to account for differences in transmission and distribution pipelines, such as differences in pipeline size and operating pressure.[2] Some stakeholders have suggested that the requirements for distribution pipelines should apply to low-stress transmission pipelines because of the similarity in operating pressures of these two types of pipelines. You asked that we consider the safety implications to the public of this proposal. Accordingly, this report focuses on how applying distribution integrity management requirements to low-stress gas transmission pipelines might affect the safety of these pipelines.

To perform our work, we reviewed and compared laws, regulations, and guidance from PHMSA. We also interviewed a broad range of stakeholders, including representatives of industry trade associations, pipeline safety advocacy groups, state pipeline agencies, and pipeline operators. In addition, we surveyed 52 officials from state pipeline safety agencies[3] in 50 states and the District of Columbia— achieving a 100 percent response rate—to collect information otherwise not available from PHMSA, including state officials' views on changing safety requirements for low-stress transmission pipelines. We also analyzed technical documents examining the point at which pipelines leak or rupture, reports on the integrity management program and related progress, and pipeline accident reports. Furthermore, we visited state pipeline safety officials and pipeline operators and associations in Colorado, Texas, and Pennsylvania, where we interviewed officials and representatives to obtain firsthand information on pipeline safety issues— including the potential effects of applying the new requirements for distribution pipelines to low-stress gas transmission pipelines—and examined pipeline infrastructure. We selected these sites as illustrative of differences in overall pipeline mileage, geography, and emerging pipeline issues.

We conducted this performance audit from February 2011 to January 2012 in accordance with generally accepted government auditing standards. Those standards require that we plan and perform the audit to obtain sufficient, appropriate evidence to provide a reasonable basis for our findings and conclusions based on our audit objectives. We believe that the evidence obtained provides a reasonable basis for our findings and conclusions based on our audit objectives.

SUMMARY

Applying PHMSA's new distribution integrity management requirements to low-stress transmission pipelines would result in less prescriptive safety requirements for these pipelines. Overall, requirements for distribution pipelines are less prescriptive than requirements for transmission pipelines in part because the former operate at lower pressure and pose lower risks in general than the latter. For example, the integrity management regulations for transmission pipelines allow three types of in-depth physical inspections.4 In contrast, distribution pipeline operators can customize their integrity management programs to the complexity of their systems, including using a broader range of methods for physical inspection. While PHMSA officials stated that "less prescriptive" does not necessarily mean less safe, they also stated that distribution integrity management requirements for distribution pipelines can be more difficult to enforce than integrity management requirements for transmission pipelines. Currently, PHMSA can grant special permits to modify requirements for individual pipelines, if merited,5 but applying the new distribution integrity management requirements to low-stress transmission pipelines would affect all such pipelines.

If PHMSA's requirements for low-stress transmission pipelines changed, operators whose systems consist almost entirely of distribution pipelines and include only a short low-stress transmission pipeline segment could benefit because they would be subject to one set of integrity management requirements instead of two. This might allow them to apply more resources to other safety priorities. However, the effect of such a change on pipeline safety is unclear. While the consequences of a low-stress transmission pipeline failure are generally not severe because these pipelines are more likely to leak gradually rather than rupture, the point at which a gas pipeline fails by rupture is uncertain and depends on a number of factors in addition to pressure. For example, the size or type of defect and the materials used to construct the pipeline also influence whether a pipeline leaks or ruptures. In addition, the mileage and location of pipelines that would be affected by such a regulatory change are currently unknown, although PHMSA recently changed its reporting requirements to collect such information. The concern is that because distribution pipelines are located in highly populated areas, the low-stress transmission pipelines that are connected to them could also be located in highly populated areas. Overall, officials we contacted from state pipeline safety agencies and PHMSA supported the current integrity management requirements for low-stress transmission pipelines. Specifically, about 58 percent of the officials from state pipeline safety agencies we surveyed (30 of 52), responded that the current requirements would best apply to low-stress transmission pipelines—affirming the current regulatory environment. By comparison, 10 of 52 officials were in favor of changing the regulatory environment by applying distribution integrity management requirements to low-stress transmission pipelines, and 12 of 52 officials stated no opinion on the issue. In light of the uncertain safety effects of changing safety requirements for low-stress transmission pipelines and the opinion of state pipeline safety officials we surveyed, the current regulatory approach of applying more prescriptive transmission pipeline requirements—with an option for operators to apply for a special permit—appears reasonable.

BACKGROUND

A network of transmission and distribution pipelines covering more than 2.6 million miles transport the majority of natural gas consumed in the United States. Gas transmission pipelines typically move natural gas across state lines and over long distances, from sources to communities. These pipelines are typically 12 to 42 inches in diameter and can generally operate at pressures up to 72 percent of specified minimum yield strength (SMYS).[6] By contrast, local distribution pipelines generally operate within state boundaries to receive gas from transmission pipelines and distribute it to commercial and residential end users. Local distribution pipelines typically range from less than 0.5 to more than 24 inches in diameter and operate well below 20 percent of SMYS. Connecting the long-distance transmission pipelines to the local distribution pipelines are lower stress transmission pipelines that may transport natural gas for several miles at pressures between 20 and 30 percent of SMYS.

The major causes and consequences of accidental releases of gas from pipelines differ for transmission and distribution pipelines. Corrosion—either internal or external—is one of the leading causes of releases from gas transmission pipelines. Given the high pressure of the gas as it is transported through these pipelines, failures can lead to catastrophic ruptures, releasing high volumes of gas that can ignite and explode. For distribution pipelines, in contrast, damage from excavation and other outside forces is the major cause of accidental releases. Distribution pipeline failures are more likely to involve slow leaks with limited volume because the internal gas pressure is much lower than for transmission pipelines. Whatever the cause, however, leaks can lead to gas migrating to and accumulating in buildings, potentially igniting and causing a fire—which could result in injury or death to residents. Natural gas has traveled underground along migration pathways, such as sewer lines, finding an ignition source some distance from the location of the release. Therefore, monitoring pipeline integrity is important to prevent both leaks and ruptures.

PHMSA administers a national regulatory program intended to ensure the safe transportation of natural gas and hazardous liquid by pipeline. In general, PHMSA has full responsibility for inspecting interstate pipelines and enforcing regulations pertaining to them,[7] although some states are designated as “interstate agents” to assist PHMSA.[8] PHMSA also has arrangements with the 48 contiguous states, the District of Columbia, and Puerto Rico to assist with overseeing intrastate pipelines. PHMSA and participating state pipeline safety offices oversee operators’ compliance with two types of safety requirements: minimum safety standards and a supplemental, risk-based integrity management program. The minimum safety standards generally cover the design, construction, testing, inspection, operation, and maintenance of all pipelines, but generally do not account for differences in the kinds of threats and the degrees of risk that individual pipelines face. By contrast, PHMSA’s integrity management program requires operators to periodically[9] assess their pipelines to identify threats and mitigate risks to pipeline segments in areas where the consequences of a pipeline failure would be most severe, such as populated areas.

The Pipeline Safety Improvement Act of 2002[10] required the Secretary of DOT to prescribe standards for gas transmission pipeline operators to implement integrity management plans, which led to the implementation of PHMSA's integrity management program for transmission pipelines in 2004.[11] Pursuant to the 2002 act, all gas transmission pipeline operators must periodically assess for and mitigate safety threats, such as corrosion. Integrity management requirements for transmission pipelines focus on portions of the pipeline located in highly populated or frequently used areas, like residential areas or parks, where significant consequences could result if an incident occurs. Such areas are referred to collectively as *high-consequence areas*.[12] Because the majority of transmission pipelines often traverse rural areas, including areas that are sparsely populated or where consequences would be low, the percentage of natural gas transmission pipeline mileage that is subject to integrity management requirements is small, about 7 percent. PHMSA also established provisions within transmission integrity management for low-stress transmission pipelines, in recognition that these pipelines are more likely to fail by leak rather than by rupture compared to higher pressure transmission pipelines and thus pose a reduced risk to the public. For example, an operator could conduct an electrical survey to address the threat of external corrosion, instead of utilizing the full external corrosion direct assessment process.

The Pipeline Inspection, Protection, Enforcement, and Safety Act of 2006[13] mandated that PHMSA also establish minimum standards for integrity management programs for distribution pipelines; PHMSA published these standards in a final rule in December 2009.[14] These integrity management programs are similar to those required for gas transmission pipelines, but tailored to reflect the differences in distribution pipelines. For example, all distribution pipelines are considered to be in high-consequence areas because they are largely located in populated areas. As a result, distribution integrity management requirements apply to all distribution pipelines. The rule took effect February 2, 2010, and gas distribution operators were required to implement an integrity management program no later than August 2, 2011.[15]

Commenters on the distribution integrity management final rule in 2009 suggested that distribution operators that also operate low-stress gas transmission pipelines should be able to use distribution integrity management requirements for those transmission pipelines, rather than implement a separate integrity management plan.[16] Commenters suggested that this could be done by amending the distribution integrity management rule or by changing the definition of a transmission pipeline, but recognized that additional rulemaking might be needed as a result. In the final rule, PHMSA noted that stakeholder groups that studied the appropriateness of such a change concluded that additional technical work would be needed to support it. Specifically, they said a better understanding of the threshold between leakage and rupture was needed to determine if low-stress transmission pipelines should be regulated under a distribution integrity management program. PHMSA concluded that it might be appropriate to consider the change at a later date, and agreed with the need for additional technical work to inform any decision.

Less Prescriptive Requirements for Low-Stress Transmission Pipelines Could Benefit Some Operators, but Safety Effects Are Unknown

Changing Safety Requirements Would Result in Less Prescriptive Regulations for Low-Stress Gas Transmission Pipelines

Overall, integrity management—and most other—pipeline safety requirements are less prescriptive for distribution pipelines than for transmission pipelines. PHMSA's integrity management requirement for assessments of gas transmission pipelines illustrates the prescriptive nature of the safety requirements for transmission pipelines. This requirement directs operators to perform an integrity assessment (physical inspection) of their transmission pipelines using a specific method[17]—in-line inspection,18 hydrostatic pressure testing,19 or direct assessment.20 Both in-line inspection and hydrostatic testing involve tools or techniques applied inside the pipeline, while direct assessment tools and techniques are applied externally. These assessment methods allow operators to detect specific anomalies21 in the pipeline, such as imperfections in the pipe wall or weld, that could lead to failure. (See fig. 1 for an example of the equipment used for hydrostatic testing and fig. 2 for an example of the equipment involved in in-line inspections.)

In developing integrity management requirements for distribution pipelines, PHMSA did not prescribe the same types of assessments as it did for its transmission pipelines, because the assessments would often not be practicable for distribution pipelines. For example:

- Distribution pipelines are typically too narrow in diameter to accommodate in-line inspection tools.

- Approximately half of the distribution pipeline system is nonmetallic, meaning that internal inspection tools cannot be used.[22]
- Hydrostatic testing and direct assessment are time consuming and can be risky and inconvenient for the public. During a hydrotest, pipelines must be shut down and gas deliveries must be stopped or curtailed for some time.
- Hydrostatic testing and direct assessment can interfere with normal activities in the vicinity of the testing. In particular, direct assessment involves excavation and disturbs property and infrastructure, potentially including roads and other utilities, with which distribution pipelines may be integrated.

PHMSA's approach for distribution pipelines requires all operators to implement an integrity management program that sets high-level performance objectives with implementation guidelines. For example, PHMSA requires pipeline operators to develop an integrity management program incorporating certain elements—such as identifying risks, evaluating and ranking risks, and identifying measures to address risks.[23]

Source: Texas Oil and Gas Association.

Figure 1. Hydrostatic Testing Equipment Used to Test Pipeline Integrity.

Source: Texas Oil and Gas Association.

Figure 2. Example of In-Line Inspection Equipment to Detect and Record Pipeline Anomalies.

Operators have the flexibility, however, to create a plan at their discretion as long as they demonstrate how it satisfies integrity management requirements, which could include a broad range of preventative or mitigative methods. For example, a distribution operator may conduct a guided wave pipeline assessment, which uses ultrasonic waves to scan and inspect the pipe. While PHMSA officials stated that "less prescriptive" does not necessarily mean less safe, they also stated that integrity management requirements for distribution pipelines can be more difficult to enforce than integrity management requirements for transmission pipelines. This is because transmission pipelines must meet a specific set of fixed requirements as opposed to the flexible program allowed under distribution integrity management.

Other regulatory requirements—not just integrity management regulations—are also generally less prescriptive for distribution pipelines than for transmission pipelines, because distribution pipelines operate at lower pressures and pose lower risks. Specifically, regulations for valve spacing, construction, patrolling, line markers, recordkeeping, and leak repairs are all less prescriptive for distribution pipelines. For example, operators must patrol distribution pipelines to observe surface conditions at least twice a calendar year in areas or on structures where physical movement or external loading could cause failure or leakage, but for transmission pipelines in the most densely populated (high-consequence) areas, operators must patrol at least four times per year. As a result, if low-stress transmission pipelines adopted distribution pipeline requirements, then low-stress transmission pipelines located in highly populated (high-consequence) areas would be subject to less frequent leak surveys and patrolling than other transmission pipelines.

As an alternative to changing the safety requirements for all low-stress transmission pipelines, PHMSA can currently issue special permits exempting. The special permit process would allow operators from compliance with one or more regulations.24 PHMSA to evaluate the safety implications of the exemptions on a case-by-case basis, as opposed to applying less prescriptive requirements across the board for these types of pipelines. As of January 2012, PHMSA officials told us they have not received any requests for special permits to apply distribution integrity management requirements to low-stress transmission pipelines.

Changing Requirements Could Benefit Some Operators, but Overall Effects on Safety Are Unknown

Changing the safety requirements for low-stress gas transmission pipelines could simplify compliance and create efficiencies for some pipeline operators, potentially allowing them to apply resources to other safety priorities. In 2006, we reported that according to operators, integrity management can be costly, when, for example, additional staff or contractors must be hired to implement integrity management requirements.[25] Under the current regulations, some pipeline companies operating both distribution and low-stress transmission pipelines must address requirements for both distribution and transmission integrity management programs. While it is possible to have one integrity management plan that addresses both distribution and transmission integrity management requirements, some of these companies operate only a small segment of transmission pipeline that connects to an entire distribution pipeline network. For such companies, changing the safety requirements for low-stress transmission pipelines could create efficiencies because they would be subject to just one set of integrity management requirements. Furthermore, operators devote resources to comply with pipeline safety regulations, and these safety resources could be made available for other safety purposes.

While changing the requirements for low-stress pipelines would benefit some operators, the impact on pipeline safety is unclear. For example, although low-stress gas transmission pipelines typically operate at pressures at which a pipeline failure will result in a leak, recent research has indicated that there is no definitive threshold at which a pipeline will leak as opposed to rupture, since pipeline integrity depends on a number of characteristics in addition to operating pressure. Though in general the threshold is in the range of 20 to 30 percent of SMYS—which is the typical operating pressure for low-stress transmission pipelines—the point at which a pipeline will rupture instead of leak varies depending on factors such as toughness, wall thickness, and the diameter of the pipe segment. For example, a 2011 study[26] determined that the boundary could range from slightly below 20 percent of SMYS for pipelines with rarely used pipe materials to well over 30 percent of SMYS for many other pipelines with more robust materials. Furthermore, a 2001 study[27] determined that pipelines operating at less than 30 percent of SMYS are more likely to leak than to rupture, but noted that pipelines operating at lower pressure have also ruptured, because of a combination of factors such as the length of the defect and toughness of the pipe. According to one of the researchers of the study, additional factors—such as damage incurred during installation and subsequent pressure increases around the damage—could increase the likelihood of an incident, even for lower pressure pipelines. In addition, some high-profile incidents have occurred on transmission pipelines operating at pressures slightly higher than 30 percent of SMYS.[28] For example, on September 9, 2010, a 30-inch-diameter underground natural gas transmission pipeline operating at 35 percent of SMYS in a residential area of San Bruno, California, suddenly ruptured. The resulting explosion and fire killed 8 people and destroyed 38 homes.

Another reason why the effects of changing the safety requirements for low-stress transmission are unclear is that the mileage and location of pipelines that would be affected by such a change are currently unknown. The concern is that low-stress transmission pipelines subject to a change in safety requirements could be located in high-consequence areas. PHMSA did not collect this information in the past, although it recently revised its reporting requirements to do so.

PHMSA's transmission pipeline annual report form—revised in November 2010 for 2011 reporting[29]—now requests information on the number of miles of pipeline and location by operating percentage of SMYS for each operator. Furthermore, of the 52 officials from state pipeline safety agencies we surveyed, 31 responded that they do not track the location of the low-stress gas transmission pipelines in their state.[30] However, by definition, low-stress transmission pipelines could be located in high-consequence areas because these pipelines connect to distribution pipeline networks. Therefore, a change in safety requirements could mean that low-stress gas transmission pipelines in high-consequence areas would now be subject to less prescriptive integrity management requirements.

Overall, officials we contacted from state pipeline safety agencies and PHMSA supported the current integrity management requirements for low-stress transmission pipelines. For example, of the 52 officials from state pipeline safety agencies we surveyed, 30 responded that transmission pipeline integrity management requirements would better apply to low-stress transmission pipelines— affirming the current regulatory environment. By contrast, 10 of 52 officials from state pipeline safety agencies responded that distribution integrity management requirements would be better for these types of pipelines.[31]

Furthermore, according to some PHMSA officials, lower stress transmission pipelines should be subject to more prescriptive requirements than would be applicable under the distribution integrity management program because low-pressure leaks can still cause significant damage.

CONCLUSION

Even though some pipeline operators might benefit from changing the regulations for low-stress transmission pipelines, the uncertain safety effects raise concerns about this change. In the current regulations, pipelines operating at below 20 percent of SMYS are subject to less prescriptive requirements, but a change could mean that pipelines operating at up to 30 percent of SMYS would also be subject to less prescriptive requirements. Given the lack of a definitive threshold at which pipelines that operate in this pressure range will rupture, not knowing the mileage and location of pipelines that would be affected by a change in requirements and not knowing the potential proximity of these pipelines to high-consequence areas makes such a change questionable.

Furthermore, the majority of state pipeline safety officials responding to our survey support maintaining the current regulatory environment, which includes an option for operators to apply for a special permit if complying with requirements for two separate integrity management programs would be burdensome.

AGENCY COMMENTS

We provided a draft of this correspondence to DOT officials for their review and comment. The officials provided technical corrections which we have incorporated throughout.

Susan Fleming
Director,
Physical Infrastructure Issues

INDEX

D

E

F

G

H

I

J

K

L

M

N

O

U

V

W

Y